WHAT EVOLUTION LEARNS AND OTHER ESSAYS

WHAT EVOLUTION LEARNS AND OTHER ESSAYS

Steven Bratman, MD, MPH

SPONTANEOUS ORDER PUBLICATIONS

Spontaneous Order Publications
Albany, NY. USA

Copyright © 2024 by Steven Bratman
All rights reserved. No part of this publication may be reproduced, distributed, or transmitted in any form or by any means, including photocopying, recording, digital scanning, or other electronic or mechanical methods, without the prior written permission of the publisher, except in the case of brief quotations embodied in critical reviews and certain other noncommercial uses permitted by copyright law. For permission requests, please address Steven Bratman at stevenbratman@gmail.com

Set in Baskerville Regular, with highlights in Century Gothic Pro

Printed in the United States of America

Print ISBN: 9798339272649
Library of Congress number on record.

The cover image shows an ice cast that models the form of the branch it once fully surrounded before sunshine melted its upper surface away.

ACKNOWLEDGMENTS

Tobias Uller and Richard A. Watson graciously answered my many questions on developmental memory, gene regulatory networks, Hebbian optimization and natural induction, and went so far as to read and provide exceedingly helpful critiques of the piece as a whole.

Richard Fikes read and gave me both encouragement and useful feedback on two essays in this collection.

Rollin Kennedy's editorial contributions were invaluable.

CONTENTS

Acknowledgments v
Introduction 1
WHAT EVOLUTION LEARNS 5
   What Evolution Learns 11
      Monkeys Typing *Hamlet* 13
      A Game of Guess and Check 17
      An Expanding Theory 21
      A Computational View 24
         Biased Variation 30
      Developmental Bias Facilitates Evolution 32
      Evolutionary Design Motifs 36
      Eukaryotic Gene Regulatory Networks as
         Biological Computers 37
      GRN Outputs: Proteins 43
      Mutations Reprogram GRNs 44
      Modularity 49
      Weak Linkage or Pluggable Modularity 53
      Modular Eyes 55
      Growing a Multicellular Organism 58
      Attractors 61
      Compartmentalization and the Universal
         Animal Toolkit 64
      Switches and Dials 66
      Exploratory and Adaptive Processes 72
      Skeleton-Key Widgets 78
      How Has Evolution Found Its Skeleton-Key Widgets? . . 84
      A Brief Introduction to Induction,
         Generalization, Extrapolation and Modeling 86
      Optimization of Optimization, of Optimization 90
      Three Primary Timescales of Biological Optimization . 93
      Optimization Timescale One: Phenotypic
         Adaptation/Plasticity 95

Optimization on Timescale One Guides
    Optimization on Timescale Two via
    the Baldwin Effect: Phenotype-First, or
    Genes as Followers 99
Optimization Timescale Three: Hebbian
    Relationships and Modularity 110
    What Watson Recognized........................ 112
    Associative Learning 114
    How Developmental Processes Build
        Hebbian Relationships 118
    Hebbian Relationships Model the
        Correlational Structure of the World 124
    The Origin of Modularity in Hierarchical
        Hebbian Relationships 125
Optimization on Timescale Three Facilitates
    Optimization on Timescale Two
        Through Developmental/Evolutionary Memory . 128
    Whales, Frogs and Snakes........................ 129
    The Persistence of Memory...................... 132
    The Hypothetical Evolution of Land Dolphins........ 135
    Developmental Memory and Expedited
        Mutational Pathways 137
    Developmental/Evolutionary Memory
        Facilitates Variation 140
Optimization on Timescale Three Guides
    Optimization on Timescale Two by
        Modeling the World........................ 141
    Brief Intuitive Explanation........................ 146
    Love the One You're With 148
    Natural Induction Through Self-Modeling in
        a Physical System........................... 156
    More Detailed Intuitive Explanation 161
    Scales and Evolutionary Moments................... 167
    Hebbian Generalization vs. Other Forms of
        Facilitated Variation 169
    Is Watson's Theory True?......................... 170
Implications for Global Climate Change............. 177
Epilogue: The Evolution of Evolvability............... 180
Major References 185
Glossary.. 186

THE NONLINEAR WORLD OF YOSHITSUGO OONO 189
The Nonlinear World of Yoshitsugo Oono 193
One-Gram Masses on Distant Stars................... 197

- Models 199
- Phenomenology and Understanding the World 204
- Theories of Mind 206
- Isolated Systems 208
- The Realm of the Densely Packed 210
- Scale Interference 212
- Linear vs. Nonlinear Effects 213
- Chaotic Systems 215
- The Almighty Noise 217
- Averaging 220
- Standing Back 221
- Special Preparations of Matter 222
- Meditation on an Acorn 224
- Knowledge 225
- Value, Morality and Meaning 228
- The Origin of Life 230
- Can Physics Successfully Address the Origin of Life? 235

WHY ROBOTS CRY 237

Why Robots Cry 241
- A Robot's First Essay 248
  - Sentience 250
  - Consciousness 252
  - Self-Awareness 253
  - Inwardness 254
  - Love 255
  - What Stimulates the Sensation of Q? 256
  - Can a Sufficiently Complex System Suddenly Emerge into Consciousness? 257
- What About Me? 258

DOUBLE-BLIND STUDIES AND THE PROBLEM OF TRUTH . 259

Double-Blind Studies and the Problem of Truth 263
- What Is a Double-Blind Study? 263
- Why Double-Blind Studies? 263
- The Rogue's Gallery of Confounding Factors 264
- The Placebo Effect 265
- Beyond the Placebo Effect 267
- Statistical Illusions 269
- Observational Studies 270
- Double-Blind Studies and Nothing but Double-Blind Studies 272
- What About When Double-Blind Studies Are Impossible? 273

Endnotes
.... 277

INTRODUCTION

*"The answer that uproots the question
from its ground is truly inspired."*

INAYAT KHAN.[1]

This is the third book in a series on cutting-edge scientific ideas that remain little-known to the public. These paradigms and theories are not yet fully developed and are unlikely to be correct in every detail, but I believe they ask the right questions and outline the right ways of thinking about the difficult subjects they address.

The first of these books, *Spontaneous Order and the Origin of Life*, explains in less technical terms the theory of life's origins presented by physicists Eric Smith and Harold Morowitz in their monumental *The Origin and Nature of Life on Earth: The Emergence of the Fourth Geosphere*.[2] Most speculation on the emergence of life investigates how amino acids and other biological molecules could have formed on early Earth. Smith and Morowitz transcend such inquiries by shifting the level of abstraction. Instead of exploring the origin of particular biomolecules or living organisms, they study the biosphere as a whole, as if viewing it from the Moon. From this more distant perspective, Earth's biosphere appears as a novel state of planetary matter existing alongside the more typical matter states that form Earth's hydrosphere, lithosphere and atmosphere.

Smith and Morowitz ask what actions the biosphere performs and answer that it provides a channel for a flow of energy that would not exist on a lifeless Earth. They see its origin as a sequence of phase transitions, spontaneous rearrangement processes that progressively facilitated the movement of electrons from higher to lower energy levels. The early Earth, they say, *relaxed* into the biosphere.

The second book in the series, *Cooperation and the Evolution of Human Nature*, largely follows philosopher of life sciences Kim Sterelny and comparative psychologist Michael Tomasello to suggest that the game-theoretic situation faced by human ancestors caused them to spontaneously evolve an "operating system" for obligatory cooperation.

Advanced cooperation is absent in apes and rare in any animal other than the various social insects, but humans depend on it. Few of us could survive alone. The daily texture of human life consists of a consistent exchange of objects and services with others, and it was much the same for our ancestors going back at least a million years. Not only do we provision one another with material necessities, we also cooperate at the level of culture, and here, too, we show the limits of our individuality: If it were possible to quantify the contents of our minds, only a small percentage would be found to belong to ourselves alone. Humans occupy a niche partway between solitary creatures like mountain lions and the near loss of individual identity seen in ants and bees.

Cooperation is widespread in biology, but all attempts to explain its emergence must contend with game theory problems in the family of the prisoner's dilemma. Simply put, organisms that evolve tendencies toward altruistic tendencies typically lose fitness by doing so and produce fewer surviving offspring, thereby weeding themselves out of the population. Peter Richerson, Robert Boyd and other authors have offered subtle solutions to this dilemma based on group selection; to oversimplify their thesis, cultural groups that cooperate well defeat groups that cooperate poorly. Against this, Sterelny argues persuasively that known facts of human origins rule out any significant contribution from group selection until quite late in our evolution, and perhaps not even then. He offers a straightforward path toward advanced cooperation, beginning with forms of direct, immediate-return mutualism that acted as a nucleus or kernel around which more advanced forms of cooperation crystallized. Once understood, his proposal seems more likely than any alternative.

The extended title essay of the book you are currently reading, *What Evolution Learns*, explores what might be called the evolution of evolvability. A close study of the deep structure of evolution reveals that Darwinian optimization processes have acted upon themselves to facilitate future evolution. Marc Kirschner and John Gerhart have offered a set of specific explanations for the remarkably rapid evolution of animals and plants in their theory of facilitated variation. Computer scientist Richard A. Watson and his research associates go further by identifying a previously unrecognized spontaneous optimization process they call natural induction that shapes the production of evolutionary variation in useful ways. Certain characteristics of evolution, they show, bear a profound, specific and hitherto unrecognized similarity to neural network learning systems. In effect, evolution has learned to generalize and uses that capacity to bias variation in adaptive directions. This nearly book-length essay is the first presentation of these paradigm-shifting ideas at a non-specialist level.

Both gene regulatory networks and the brain's neural networks operate by constructing models of the world. The second essay, based on the fertile, surprising and sometimes head-spinning ideas of physicist Yoshitsugo Oono, attempts to answer a deep question: How is it that the world can be modeled at all? The third essay builds on elements of the first two to propose a deflationary account of such metaphysical concepts as consciousness, sentience and self-awareness. The final essay shows how our intuitive models of the world can easily go astray, as demonstrated by the counterintuitive results of double-blind studies.

WHAT EVOLUTION LEARNS

In their 2006 book *The Plausibility of Life*,* evolutionary biologists Marc Kirschner and John Gerhart outlined a collection of evolutionary and developmental mechanisms that make evolutionary innovation more efficient. This essay began as an exploration of their ideas. However, while researching the subject, I encountered a paper by Tobias Uller and colleagues that caused me to see the subject differently (Uller et al., 2018). In 12 compelling pages, Uller reconceptualizes evolution as a progressive transformation of gene regulatory networks—the DNA-hosted computational systems that determine where and when genes are expressed. These systems bias the production of variations and, by so doing, enhance evolvability.

An intriguing sidebar in the Uller paper, as well as several citations in the paper's body, led me to the learning systems approach of Richard A. Watson. That material was another revelation. Evolution has long been analogized to learning, but Watson has converted those vague analogies into something closer to equivalence.

---

* This book and all the other primary references cited in this essay are listed in the Major References section.

Evolution is a spontaneous adaptive process that finds solutions to environmental challenges through slow processes of brute-force trial and error. Brains also discover adaptive responses to challenges but do so much more intelligently. Creatures equipped with brains surpass mere trial and error because they learn as they go along and acquire the ability to generalize from past experiences.

But are these two forms of adaptive problem-solving completely different? Newly emerging insights suggest otherwise. Watson and other researchers in the field of computational evolutionary biology have found deep similarities between long-term evolutionary processes and the neural network learning processes that occur in brains and artificial neural network machine-learning systems. Their work suggests that evolution, too, has moved past mere trial and error and can now make intelligent guesses.

Evolution proceeds not only through mutations in genes but also through shifts in the systems that regulate gene expression. Watson's analysis shows that these "gene regulatory networks" transform over evolutionary time in much the same way that the neural networks in our brains transform as we learn through life experience. In the language of artificial neural network machine-learning systems, evolving gene regulatory networks undergo a form of unsupervised learning. When faced with novel environmental challenges—such as anthropogenic climate change—populations of organisms can effectively generalize from past adaptations to "propose" novel variations that have an elevated likelihood of working out. Evolvability has itself evolved.

Watson's ideas remain a work in progress, but even at their current state of development, they are persuasive and intellectually thrilling. His perspectives shape this entire essay.

If I were willing to write in the hyperbolic style of popular science, I might say something like the following:

> Once you have read this, you will never look at evolution in the same way again! You will see that

there is more to it than the ad hoc emergence of random traits that emerge through chance historical processes and undergo selection. Instead, you will come to view evolution as a process of constant learning. Natural selection operating on heritable variation has fed back on itself to produce a DNA-hosted learning system that resembles nothing so much as the deep-learning neural networks that power AI systems like ChatGPT; however, instead of modeling language, this system models Earth environments and how to survive within them! The heritable variation that natural selection operates upon was once random and directionless, but today, it is shaped by evolved models written into developmental systems that output variations intelligently. Evolution has learned how to evolve!

But I am uncomfortable with writing that permits itself exclamation marks and polishes to a golden brightness what is already wonderful in its unpolished state. I will claim only this: Watson's theory is sufficiently plausible that it is worth taking seriously as a guide to future research. Those scientists who study the intricate circuits of gene regulatory networks and how they operate to assemble traits from sub-traits and sub-sub-traits might find it rewarding to look for evidence of neural network-like modeling and generalization. In addition, Watson's paradigm may enhance our understanding of complex systems generally.

And yet—for I am not unmoved—I am compelled to add that these ideas are magnificent, mind-bending, deeply fascinating, potentially profound and perhaps revolutionary.

WHAT EVOLUTION LEARNS

"All wisdom comes from memory."

AESCHYLUS, *PROMETHEUS BOUND*.

Darwin's theory is often described as "survival of the fittest," but that description leaves out half the story. There can be no natural selection without variation—the differences in form and function that cause one organism to survive and reproduce more successfully than another. But what produces the variation that selection acts upon?

The usual answer is genetic mutation, DNA changes that randomly modify the structure of proteins. Through endless iteration, these minute, directionless alterations add up to produce the "endless forms most beautiful" that populate the Earth.* However, it is not immediately obvious that random alterations to protein structures should be able to

---

* From *The Origin of Species*: "It is interesting to contemplate an entangled bank, clothed with many plants of many kinds, with birds singing on the bushes, with various insects flitting about, and with worms crawling through the damp earth, and to reflect that these elaborately constructed forms ... have all been produced by laws acting around us. ... There is grandeur in this view of life, with its several powers, having been originally breathed into a few forms or into one; and that, whilst this planet has gone cycling on according to the fixed law of gravity, from so simple a beginning endless forms most beautiful and most wonderful have been, and are being, evolved."

serve up responsive, useful, creative variations rather than meaningless shifts that go nowhere and terminate at dead ends.

If asked how evolution manages to forge ahead so successfully, evolutionary biologists have traditionally pointed to the power of time and numbers. The global population of bacteria has been estimated at 100 trillion *trillion*, and each one reproduces hourly. When carried out on such a monumental scale over billions of years, mere trial and error seems capable of accomplishing great feats of innovation.

But Darwin had animals and plants in mind, not bacteria, and animals neither come in the trillions of trillions nor reproduce 24 times a day. If meaningful variation depends entirely on raw numbers, evolution should have downshifted when it came to animals. However, that doesn't seem to describe what has happened. Bacteria have been around for several billion years, and they are masters of chemistry, but animals are something else again. Fish transformed into the first amphibians in 160 million years; amphibians became dinosaurs in another 140 million; and within a scant 80 million more, dinosaurs grew feathers and took flight. Human ancestors more than tripled their brain size in only three million years as they differentiated from other apes. Rather than losing steam, evolution seems to have accelerated. Animals (and plants; probably fungi, too) seem to possess accelerated creative powers, enhanced capacities to innovate adaptive variations that survive the test of selection.

In human life, we use the adjective "wise" to describe individuals whose rich past experiences and habits of reflective understanding have granted them insight into the patterns, trends and recurrent tendencies that lie beneath the superficial variety of human events. When novel challenges arise, people with wisdom can generalize from their past experiences and offer informed advice on how best to respond. Their advice may prove incorrect, but it is more likely to be on the right track than the suggestions of those who lack the same experiential depth.

As we shall see, evolutionary processes, too, learn from experience, and over billions of years of finding successful adaptations, they have "learned" the deep patterns of changing Earth conditions. This knowledge or wisdom is not remembered in brains but is physically embodied in the developmental systems that translate mutations in DNA into changes in organism form. While variations were once random, they now emerge through the operation of a mechanism that "knows" a great deal about what adaptations are worth trying in the face of current challenges. This deep, embedded knowledge has progressively freed evolution from dependence on brute-force trial-and-error and given it the ability to make what amounts to intelligent guesses.

Or, at least, that is the thesis of this essay.

Monkeys Typing *Hamlet*

Fred Hoyle was a brilliant physicist of the mid-20th century who reflected on evolution and concluded that Darwin must be wrong. He wrote, "The chance that higher life forms might have emerged in this way is comparable with the chance that a tornado sweeping through a junkyard might assemble a Boeing 747 from the materials therein."[3]

Elsewhere, Hoyle compared the possibility of the emergence of life through random events to the familiar image of a roomful of capuchin monkeys typing out the text of *Hamlet*. Simple calculations show that a trillion, trillion, trillion monkeys couldn't possibly type *Hamlet* even if given a trillion, trillion, trillion years to try. And *Hamlet* contains substantially fewer data elements than the protein-coding DNA of simple bacteria. These overwhelming numbers made Hoyle a lifelong opponent of Darwin's theory.

But Hoyle got it wrong, and in more than one way. Evolutionary biologist Richard Dawkins famously blasted back at Hoyle's analysis in his 1976 book *The Selfish Gene*.[4] I will present a version of Dawkins' famous refutation in the next section. Here, I offer a different argument based on traits and learning processes.

Hoyle had missed the fact that the world of living things—the biosphere—learns as it goes along.

Ignoring punctuation and spaces, the text of *Hamlet* contains about 130,000 characters.[5] There are approximately $10^{184,000}$ possible character sequences of that length, and to have a 95 percent chance of randomly typing out *Hamlet*, monkeys would have to produce about $3 \times 10^{184,000}$ *Hamlet*-length texts. If a trillion, trillion, trillion monkeys typed away at one second per character, this would take them about $10^{183,969}$ years. Given that the universe is only about 10^{10} years old, this presents a problem—and if one substitutes typical bacterial DNA for *Hamlet*, the problem only gets worse.

Hoyle's analogy treats each organism as if it had no history and came from nowhere, but that's not at all how evolution works. Organisms always emerge from prior organisms and build on their past successes. To see why this makes such a difference, let us return to our roomful of monkeys and give them the ability to learn.

Actually, we will need only one monkey. Suppose a single hard-working capuchin sequentially types out 130,000-character documents and, after completing each one, shows it to a supervising human who marks each correct letter with a red checkmark. Our monkey is a good student. When it produces its next 130,000-character rant, it takes care to keep the red-checked letters intact and only sets its monkey madness free when typing the remaining letters. Using this system, the monkey needs only type 376 documents to have a 95 percent chance of typing *Hamlet* flawlessly.[6] If it can type 200 characters per minute, a mediocre rate for humans, and does so for 12 hours each day, it will hammer out the play verbatim in less than a year.

The above is not an accurate description of how evolution operates at the level of DNA, but it bears a meaningful similarity to evolutionary processes at the level of organisms and their adaptations. Evolution learns as it goes along and tends to remember what it has learned.

Note: When I write that evolution "does" X, I am using a metaphorical shorthand for "the totality of evolutionary processes yield X." I certainly do not mean to imply that "evolution" is an agent or an entity. The information that evolution "possesses," and that at every evolutionary moment it "uses" to "build" organisms, has been encoded into the physical structures of past and current organisms through purely mechanical algorithmic processes. But those algorithms have certainly produced remarkable results.

As organisms evolve, they achieve successful *adaptations*: the many and various systems, structures and processes that enable organisms to survive and reproduce. Some adaptations are specialized and transient, but many are sufficiently general-purpose in their capacities and well-suited to their purposes that they endure with only limited change for extended periods. Once evolution has learned how to do something well, it typically holds on to that knowledge, more commonly tweaking and modifying successful adaptations than fully abandoning them.

For example, the systems that manage core metabolism in aerobic organisms have remained much the same for several billion years, as have those that transcribe DNA into RNA and translate RNA into proteins. These were major evolutionary inventions that, once established, were never abandoned. Other highly conserved processes include the use of ATP for cellular energy use and storage, hemoglobin for oxygen transport and the various homeobox genes that play a role in all eukaryotes and control developmental processes in animals. More recent near-universal adaptations specific to vertebrates include the use of the protein keratin for protection of external surfaces, bones for rigid structures and a set of four jointed legs for locomotion on land.

Over evolutionary time, organisms have progressively acquired useful tricks, methods and techniques that allow them to survive and thrive. These (largely) settled adaptations are dependable solutions to various problems of survival on Earth, and they are the (loose) equivalent of red-checked letters in the monkey's successive texts. The equivalent of typing *Hamlet* is producing a successful organism in the current environment.

What Hoyle missed is that organisms—and airplanes—are not created from scratch but grow out of past discoveries.

Our protohuman ancestors would have had about the same chance as a tornado of assembling a 747 from the contents of a junkyard, and even the Wright brothers could not have done much better, especially if they had been confined to working with junkyards as they existed in the early 1900s. The ability to build 747s depends on the knowledge previously acquired through building biplanes and the construction of biplanes on technologies invented still earlier, including the internal combustion engine and the screw propeller. At each moment of human technological development, some aspects of knowledge are in flux, but a vast pyramid of settled knowledge lies beneath. Every physical part and all the fundamental technologies used to build a 747 depended on prior devices and technologies painstakingly worked out by striving humans in an unbroken chain back to the first stone tools.

Much the same is true of evolution. The evolutionary invention of legs depended on the prior invention of fins and fins on the invention of cartilage, bones and muscle. Each characteristic, trait and adaptation found in an organism today is the product of a long process of gradual discovery, and many of the physiological processes and anatomic structures currently in use were initially innovated billions or at least hundreds of millions of years ago. Evolutionary processes have learned to build complex organisms one invention at a time.

Importantly, learning is more than simple accumulation—it also operates *on itself*. As children progress through school, they not only memorize facts but also learn how to learn. Wilbur and Orville Wright largely depended on trial and error when they built their flying machines, but modern aircraft engineers use mathematical modeling. Humans have always tried to understand the world and, over time, have learned to do so more effectively; today, we call the knowledge of how to acquire knowledge "science."

Evolution, too, has increased its capacities recursively. By the time animals came around, evolutionary processes had not only produced successful adaptations to numerous challenges but also sophisticated models of *how* to produce successful adaptations. Evolution may have once relied entirely on trial and error, but now it makes educated guesses.

A Game of Guess and Check

One way to solve a mathematical equation is to hazard a guess and check to see if it's correct. To find the square root of 16, you can successively try two, then three, then four. Evolution, too, is a game of guess and check; the guesses are mutations and the checks are relative fitness. To improve results in a game of guess and check, you can either improve the guessing function or the checking function. In *The Selfish Gene*, Richard Dawkins showed that Hoyle's argument against the plausibility of evolution implicitly relied on an absurdly inefficient checking function. The checking function that evolution actually uses is much more effective.

The following is a simplified version of Dawkins' argument.

Suppose I have hidden an unknown object somewhere in a room and given a young child named Georgina the task of finding it while I watch. Hoyle's method limits me to giving her a thumbs-up or thumbs-down whenever she points to a stuffed bear or a pencil. But that's a lousy way to search for anything and Georgina would have to keep at it all afternoon. If evolution, too, were limited to a success/fail checking function, Hoyle would be correct and nothing interesting could ever happen.

Instead, suppose that I can give Georgina sliding-scale feedback by saying, "You're getter warmer" or "You're getting colder." Each time I say "warmer," she takes one step in a random direction from where she currently stands; if I say "colder," she goes back to her previous position and takes a single step in a different direction. This incremental, iterative process will cause her position in the room to converge on the location of the hidden object, and she'll find it pretty quickly, as anyone who has played the game knows.

In evolution, the "object" to be found is "the fittest achievable organism in the current environment,' and the equivalent of Georgina's incremental steps are mutations. Relative fitness supplies the warm/cold hints; fitter organisms are warmer than less fit ones. This process is incremental because each offspring of a fitter organism starts somewhat closer to the goal than the offspring of a less fit one and begins the next stage of searching from there. By definition, the fittest organisms have the most surviving offspring, and they tend to take over in a population. Through iteration of this process over generations, a population of organisms can converge on optimal achievable fitness with surprising speed.

The warmer/colder method of searching a space of possibilities is an example of a "hill-climbing algorithm." Evolution proceeds through hill climbing, and as it does, it is said to "climb fitness peaks."

Dawkins' argument shows that the checking function actually used by evolution is much more efficient than the thumbs-up/thumbs-down feedback that Hoyle implicitly assumed. If natural selection ignored incremental improvements and only rewarded maximal fitness, evolutionary processes could no more have built organisms than a tornado could assemble a 747. But because fitness supplies warmer/colder feedback, evolutionary processes have been able to accomplish quite a lot. When playing a game of guess and check, it pays to use a good checking function.

But there's a second option: improving the guessing function.

Just as some methods of checking are better than others, so are some methods of guessing. Suppose I want to find the square root of two to five decimal places. There are good ways to do this and bad ways. One rather poor method might go as follows: I make a guess, square it, and check the result. If the figure I come up with overshoots two, I pick a somewhat larger number at random and try that one; if it undershoots, I pick a somewhat smaller one. This method can certainly be made to work, but it's not the best approach. I'll reach the correct answer much more quickly if I select my sequential guesses using Newton's method: Start with any random guess, divide two by that guess, average the result with the initial guess and use the averaged value as the *next* guess.

Is it possible that evolution's guesses are similarly well-chosen?

Evolutionary biologists have traditionally assumed that the answer is "obviously not." DNA mutations resemble a drunkard's walk: directionless at each step, equally likely to stumble to the right as to the left. They are like numbers pulled out of a hat. They can't be smart, clever, well-tuned or in any other way better than chance.

But there's a hidden error in this traditional reasoning. It turns on the meaning of "mutation."

DNA mutations are indeed (substantially) directionless. But natural selection operates on *organisms*, not DNA sequences, and directionless mutations at the level of DNA do not necessarily produce directionless changes at the level of mutated organisms. The developmental processes that build organisms have something to say about the matter, too.

At the very least, developmental processes *bias* the effects of mutations, making some more likely than others. Cats are sometimes born without a tail, but no kitten has come out of the womb with a unicorn horn. It might be the case that these developmental biases are of no significance, effectively just as directionless as DNA mutations. However, as we shall see, evidence suggests the opposite; developmental processes tend to channel the effects of directionless changes at the level of DNA toward adaptive changes at the level of organisms.

Early in the history of life, evolutionary progress may have depended entirely on lucky mutations.* But nowadays, the specific character of mutations has plausibly become less important in evolution than the action of the sophisticated developmental machines that convert modified DNA into modified organisms.

As a simplistic analogy, think of what happens when you repeatedly shake a container of mixed nuts. With each shake, the constituent nuts shift about randomly. Your shakes do not tell the nuts where to go, and yet, the result is something more interesting than uniform mixing: With repeated shakes, the larger nuts move to the top.

---

* Or maybe not. Phenomena like natural induction as described in this essay may have begun very early, perhaps even in the pre-cellular chemical origins of life.

Similarly, the machinery of embryonic development has taken on special characteristics so that directionless mutation—the equivalent of shaking—does not randomize organism form and function but changes it in ways that are structured and ordered. The modified organisms that emerge stand a (fairly) good chance of being adaptive right out of the gate, even before selection has had time to pick and choose the best ones.

Furthermore, the changes induced by DNA mutations can be sweeping rather than minute. Evolution long ago abandoned pointillism and now paints with a broad brush. It once programmed in machine language but now writes in a high-level language whose powerful libraries and subroutines do most of the work. Instead of drawing marginally changed sequential images to create animated movies, evolution now employs animation programs that interpolate frames automatically.

But the two computer-science analogies of the prior paragraph aren't quite correct because they invoke an artist or operator with will, intention and a plan, and evolution lacks those qualities. As a better analogy—and one that will prove to be more than an analogy—consider artificial intelligence (AI) systems based on machine learning.

Systems such as ChatGPT build models of human language by observing vast quantities of human text and discovering the deep patterns that lie within it. And they do so (largely) without human supervision. Such models "teach" *themselves* what words follow others, what sentences follow sentences and how entire sets of thoughts are organized into texts. When prompted to write something, ChatGPT at the very least produces words that are arranged comprehensibly.

If the hard-working capuchins of our monkey-typing story similarly increased in proficiency as they went along, their typing attempts would gradually come to consist of random words rather than random letters; after even more practice, they would compose in randomly chosen sentences that nonetheless make grammatical and semantic sense.

Evolution, too, has built a model—not of written language, but of effective ways to build successful organisms. This model "knows" how to build powerful adaptations, what adaptations go well with other adaptations and what collections of adaptations go well with other such collections. More subtly, evolution's model mirrors the world and how it typically changes. Features of the Earth's environment that vary independently are reflected in independently varying adaptations and correlated features by correlated adaptations. Evolution no longer writes in letters or even in words but, like ChatGPT or our hypothetically educated monkeys, in intelligible and relevant DNA texts.

Evolution has learned a thing or two about the world and how to produce organisms that thrive in it.

An Expanding Theory

The theory of evolution has been around for so long that one might be excused for thinking that its major elements are long settled and that ongoing research consists merely of the filling in (of an admittedly vast number) of details. But the true state of the field is much more interesting. The ever-increasing rate of accumulation of detail has done more than fill in blanks; it has transformed our understanding of the core processes that power evolution.

The algorithm of Darwinian selection has three parts: selection, variation and heritability. Selection, often simplified to "survival of the fittest," acts as a filter on populations of organisms. If organisms vary in their fitness—their average capacity to survive and reproduce—and if this variation is at least partially heritable, selection will gradually modify populations so that the average fitness of its members increases. The result is a spontaneous optimization process based on tautology; those organisms that produce the most copies of themselves appear in the greatest numbers.

Over time, this process has yielded organisms that can be said to possess their own optimizing capacities: They take what the world throws at them and, within the limits of their powers, respond appropriately. Even an amoeba "knows" to move toward nutrients and away from toxins. Organisms work hard to get on in life, and evolution grants their surviving descendants increased capacity to do so. Natural selection on heritable variation is thus a slow optimization process that operates on the faster optimization processes that it has itself built. And, as shall become apparent, this stacking of optimization on top of optimization operates on multiple levels.

Darwin observed the process of heritable variation, but he had no clue as to how traits are inherited or come to vary. Gregor Mendel's discovery of dominant and recessive inheritance provided the first hint, and when Francis Crick and James Watson discovered the molecular structure of DNA and Marshall Nirenberg cracked the genetic code, the sources of inherited variability seemed to have been solved.

DNA, scientists came to believe, is composed of genes that code for proteins, and mutations that alter those genes alter the proteins they code for. When proteins change, so do organisms. Most mutations are neutral, many are harmful, but a few marginally enhance fitness. Organisms evolve through a succession of such lucky, marginal improvements.

This understanding and its accompanying mathematical framework constitute the basis of the 20th-century understanding of evolution, sometimes called the *modern evolutionary synthesis*. From this perspective, evolution resembles a random exploration of a space of possibilities. Its remarkable creativity results from the three-part Darwinian algorithm: Genetic mutations introduce variations, selection reviews variants for fitness and surviving organisms carry on those fitness-enhancing mutations in their DNA.

However, we now know that this picture of evolution misses much that is essential.

First, in complex organisms, many and perhaps most useful mutations occur in non-protein-coding regulatory elements rather than in genes. These regulatory elements operate together as elements of a biological computer that controls when and where genes are expressed. When mutations occur in regulatory sequences, the effect more closely resembles changes in software than hardware. The next several sections will expand on this idea and explain its extensive implications.

More fundamentally, the three elements of natural selection feed back on themselves, and do so over multiple timescales. This multi-level self-interaction introduces complex recursive effects.

Self-action complicates *selection* because organisms created under the influence of selective forces modify the selective forces that subsequently act upon them. For example, organisms do not merely evolve in response to environmental conditions but also actively change and manage their environments; they then proceed to undergo further selection in the modified environment they have in part created. This two-way relationship is called "niche construction."[7] In addition, organisms respond to shifting environmental conditions by shifting their behavior and physical form, and this, too, alters the selective forces that act upon them. This phenomenon will be discussed extensively in the sections on phenotypic plasticity and the Baldwin effect.

Self-action also affects *inheritance* by shifting the units that come under selection; unicellular organisms that transmit genetic information individually subsequently evolved into multicellular organisms that transmit it collectively. Evolutionary transitions of this kind are described as shifts in the level of individuality.

Finally, self-action modifies potential *variation* by channeling random mutations through evolved processes that systematically bias their effects. Perhaps surprisingly, this bias seems to guide variations in useful directions. Biologists Marc W. Kirschner and John C. Gerhart have dubbed this evolution-enhancing effect *facilitated variation*. They write, "Rather than staggering along like a drunken sailor, evolution marches along a myriad of paved pathways, changing direction without instruction, but taking large, forceful steps and avoiding many lethal obstacles" (Kirschner and Gerhart 2005, p. 247).

The theory of facilitated variation includes several distinct principles, and I will explore most of them in this essay. In addition, I present a newly developed learning-systems approach largely attributable to Richard Watson that adds at least two additional principles and enfolds most of the "traditional" elements of facilitated variation into a single explanatory framework. Watson does not seem to have chosen a name for his theory, but I will provisionally call it *natural induction through self-modeling*.

Note: Nature is so byzantine, redundant, creative and opportunistic in its methods that to describe it accurately, one must qualify almost any definitive statement with a long list of exceptions and qualifications. To make this narrative readable, I have left out many such wrinkles. I will try to call attention to those omissions by prefixing statements with parenthetical caveats such as (almost) or (mostly). Nonetheless, a careful reader will undoubtedly find sentences and paragraphs where I have silently omitted exceptions. For a proper academic rendition of this material, see the cited academic articles and books, especially those listed in the Major References section at the end of the essay. I have also included a glossary.

Unless otherwise specified, the following discussion refers to animals. However, similar descriptions apply to plants and fungi and, with limitations, to unicellular eukaryotes such as amoebas.

A Computational View

This essay adopts a computational view of living organisms and evolutionary processes. At a minimum, this perspective supplies new, potentially fruitful metaphors and analogies. However, recent work in the field of computational biology suggests that the relationship is closer than a loose analogy; the similarities between evolutionary processes and various aspects of computer science may be profound. For example, the systems that regulate the expression of genes are easily recognizable as computational devices, and certain aspects of how those systems transform over time resemble the intermediate states of neural network machine-learning systems as they train on data.

This computational perspective will deepen over the course of this essay. Let's start with the relatively simple regulatory processes that operate in bacteria and archaea, the evolutionarily ancient unicellular organisms collectively called *prokaryotes*.

Prokaryotes possess only a single strand of DNA, of which about 90 percent consists of genes that code for proteins. Proteins are typically useless on their own and can only perform vital actions or create helpful structures when they are produced in the company of several other proteins. To ensure that interdependent proteins are produced simultaneously, prokaryotes collect the genes that code for them into contiguous functional units called operons; when one gene is expressed (converted into a protein), so are all of its "partners."

Each operon is immediately preceded by an on/off switch.* These switches are not genes because they do not code for proteins; instead, they are short DNA sequences that *control* the production of proteins. When a switch is in the OFF position, the genes in the operon it controls remain quiescent. When certain signals arrive and bind to the switch, it flips to the ON position, and gene expression begins. The DNA strand splits open at the site of activity, and all the genes in the operon are transcribed into messenger RNA. If converted into an intelligible command, the message carried by messenger RNA would read, "Build a protein based on the coded symbols I carry." Elsewhere in the cell, a molecular machine called the ribosome proceeds to do so.

Typically, the proteins produced by the genes in the operon can themselves bind to the operon's switch and turn it off. This influence acts as a simple feedback mechanism to ensure proteins are not produced in excessive quantities. But prokaryotic DNA also includes special genes that code for *transcription factors*, signaling proteins whose only function is to control on/off switches. Some of these tend

---

* To be more precise, each operon has a "promoter" that turns it on and an "operator" that turns it off, but the two together can be thought of as a single on/off switch that takes multiple inputs.

to push certain switches to the ON position, while others tend to flip those switches OFF. These transcription factors are themselves activated by other transcription factors, and all together, they constitute a network of regulatory interactions.

The entire collection of switches and interacting signals functions as a digital cybernetic system that controls the activation of prokaryotic genes. But it is fairly simple, utilizing only a limited number of interconnections. The sophistication and computational capacities of genetic regulatory systems will increase as we follow the evolutionary path toward animals and other multicellular organisms.

About 2.7 billion years ago, ancient prokaryotes gave rise to organisms of greater complexity known as *eukaryotic* cells. The space within a prokaryotic cell is relatively undifferentiated, but eukaryotes possess multiple internal compartments confined within membranes. These include the central nuclear membrane that encloses the DNA strands called chromosomes and the various other membranes that compose the boundaries of such structures as lysosomes, endoplasmic reticula and the energy-producing modules called mitochondria. Unicellular members of this group include yeasts, single-celled algae, amoebas, paramecia, the sexually transmitted disease-causing *Trichomonas vaginalis* and the bane of backpackers, *Giardia lamblia*.

Eukaryotes occupy 10 to 1,000 times more volume than prokaryotes, meaning that there are usually fewer of them per unit of volume; in addition, most varieties take considerably longer to reproduce. These differences would seem to suggest that eukaryotes should evolve more slowly than prokaryotes, but that isn't what happened. Instead, eukaryotes have gone from success to success, evolving wildly creative responses to the challenges of existence.

Prokaryotes carry out sophisticated chemistry, but eukaryotes take action.* Amoebas crawl toward or retreat from chemical stimuli

---

* The division is not absolute; prokaryotes, too, move under their own power. But eukaryotes take it to a new level.

and stretch out arms that surround their prey, behaviors enabled by a powered cytoskeleton that constantly forms and breaks apart. Many unicellular eukaryotes project slender flagella to help them swim; some shoot out defensive darts called trichocysts, while others possess well-adapted light receptors. For reasons that remain unclear and contested, most eukaryotes engage in sexual reproduction rather than simple division. All of this is managed by DNA-based computational systems that are far more complex and sophisticated than those that control gene expression in prokaryotes.

And then came another momentous change: Unicellular eukaryotes joined forces to become multicellular organisms.

Many unicellular organisms form simple colonies. For example, bacteria in our mouths coat our teeth with biofilms, colonial structures that permit them to engage in the mutually supportive, nutritionally fruitful work of burning holes in tooth enamel. However, after about 1.7 billion years of eukaryotic evolution—perhaps a billion years ago—currently unknown lineages of eukaryotes gave rise to permanent multicellular descendants that reproduce collectively. These eventually became the kingdoms of animals, plants and fungi, along with three forms of multicellular algae.

Multicellular organisms appear in even smaller numbers and reproduce still more slowly than unicellular eukaryotes. If evolution had continued to proceed through directionless incremental changes, one would expect it to have experimented with minute sponges for a billion years or more and then, perhaps, achieved a worm. Instead, animals and other multicellular creatures differentiated and transformed at an incredible rate. Mere incremental mutation operating on relatively few organisms with relatively slow reproduction rates seems wildly insufficient to have created such innovations as rigid trunks several hundred feet tall, birds and bugs that fly, caterpillars that metamorphose into butterflies, and octopuses whose skins act like video display terminals—and to have accomplished all this in only a few hundred million years.* By this point in evolution (as we shall see), life had learned a thing or two about how to innovate.

At this point in the narrative, I must bite the bullet and introduce two technical terms for which there is no appropriate expression in plain English: The information encoded into an organism's DNA is called its *genotype*, and the organism's anatomy, physiology and characteristic behavior its *phenotype*. This distinction is critical because selection operates on phenotypes, not genotypes. The fitness of an organism can't be determined by reading its genes; genes must be allowed to build an organism first, and it is the organism that goes on to run the gauntlet of selection.

I will repeat this because it is important: Natural selection acts on phenotypes, not genotypes.

Many distinct mechanisms can produce mutations, including, but not limited to, DNA copying errors and noxious influences such as radiation or chemical mutagens that directly knock about DNA base pairs. These errors have no direction and resemble noise. However, before selection can begin to operate, the effect of a mutation must be channeled through whatever systems produce phenotypes from genotypes. In the case of animals and plants, these are the embryonic developmental systems that produce an adult organism from a single embryonic cell through the iterative operation of sophisticated DNA-hosted computational processes.

The developmental mapping or conversion process that transforms genotype into phenotype is far from simple, especially in plants

---

* This is only an intuitive perception. It is certainly possible that we regard animal evolution as especially innovative because we are chauvinistic and find their adaptations more interesting than those that occur in bacteria. In any case, the argument that follows does not depend on a claim that animals are in any sense "better" at evolving than bacteria, just that the manner in which they evolve is somewhat different and perhaps, in a certain sense, more efficient.

and animals.† Among other complexities, it is neither continuous nor one-to-one. Ten mutations in a single region of DNA may produce no effect whatsoever, while an eleventh might cause the organism to grow a sixth toe. To borrow language used to describe mathematical functions, mutations that are somewhat far apart in genotype space may produce the same output in phenotype space, and some mutations that lie close together in genotype space may produce outputs that are relatively far apart in phenotype space.

The picture of evolution proceeding through small random increments implicitly assumes that genotype space *represents* phenotype space, as if organisms were digital images and mutations affected individual pixels. But that can't be right. If mutations worked that way, they would blur phenotypic images rather than sensibly modify them. Instead, they operate more like input parameters to a system that builds objects out of *features* such as spines, limbs, skin, brains and claws. This system takes random mutations as genotypic input and spits out structured, coherent phenotypic changes as output.

Such systems cannot help but introduce biases and constraints.

---

† It is also not so simple as a mapping. As stressed by Mary Jane West-Eberhard, an organism's genotype isn't the only contributor to its form. In animals and plants, the phenotype and genotype of the mother also play a major role, as does the environment in which the organism developed. Epigenetic factors also play a role. But for some purposes, the mapping metaphor suffices.

Biased Variation

> *"Natural selection cannot work with imaginary phenotypes, only those realized by developmental systems."*
>
> ULLER ET AL., P. 949.

Selection acts on the phenotypic variations that genotypic mutations produce, but the effects of mutations are limited to the possible actions of embryonic developmental systems. An airplane pilot can push and pull on a joystick to execute a loop-the-loop, but no combination of control actions can cause a plane to smile or scratch its cockpit with its wings. Similarly, mutations cannot produce arbitrary changes in phenotype, but only those that the organism-building system permits.

Some phenotypic changes are easier to produce than others. It seems to be particularly simple to adjust total body size given that dog breeders have had no trouble breeding two-pound Yorkshire Terriers and 300-pound mastiffs from the original wolf stock. On a cell-by-cell level, the difference between a Yorkie and an English Mastiff is much greater than between any dog breed and a variant with two tails or a small unicorn-style horn, and yet, it would be challenging to breed a horned or two-tailed dog. The fact that mutations are far more likely to shift total body size than produce other creative changes shows that the systems that build dogs are biased toward certain variations and away from others.

Developmental systems possess many forms of bias, some of them quite strong. For example, almost all phyla (large categories) of animals belong to the group *Bilateria*, so named for the bilateral symmetry of its members. Bilaterally symmetric fins and limbs embody a famous generalization about the world that animals live in: For every action, there is an equal but opposite reaction. If a mutation or developmental error caused an organism's fins or limbs to grow much larger on one side than the other, it would tend to run or swim in circles. To maintain their good fit with the world, members of *Bilateria* utilize redundant systems to maintain appendage symmetry, and those systems make it next to impossible for mutations to induce asymmetry in these appendages, at least at birth.*

Similarly, most animals that walk on all fours possess fore and hind legs of (approximately) equal length. Foreleg/hind leg symmetry, too, can be regarded as a generalization about Earth environments: "The world contains as many uphill segments as downhill ones." However, front-back limb symmetry is less forcefully established than bilateral symmetry, as demonstrated by the divergence of upper and lower limbs in birds, bats, kangaroos, rabbits and humans.

Another source of constraint can be described as the dead hand of the past. Once evolution has gone far enough in a given direction, it may become so committed to that direction that, in practical terms, it can't go back. The blind spot of the vertebrate eye is a famous example.

In vertebrates, the rods and cones that respond to light send out their nerve fibers in the wrong direction: forward toward the lens rather than backward toward the brain. They have to get back to the brain somehow, and they manage it by bundling together and punching through the retina. Because these tiny fibers are transparent, they don't impair vision in any location other than where they bundle and punch through, but that (small) region lacks light receptors and is a blind spot. Octopuses evolved their otherwise rather similar eyes along a different pathway, and their light receptors sprout nerve fibers in the correct direction, backward toward the brain. The thoroughly dumb system that vertebrates somehow got stuck with does little harm, but if there were ever an evolutionary value to rerouting all these nerve

---

* In some animals, meaningful asymmetry sometimes emerges *after* birth. For example, lobsters are born with symmetrical claws, but a slight handedness in usage patterns causes these to subsequently differentiate into a large crusher claw and a smaller ripper claw. Flounder are even weirder; they are born with symmetrical eyes, but one of them subsequently wanders over to join the other on one side of their flattened bodies.

fibers in the proper direction, evolution probably couldn't manage it without passing through an intermediate stage of blindness. In more technical terms, once evolution has climbed sufficiently far up a fitness peak, there may be no route back down that isn't precluded by a sharp loss of fitness.

Given all of the above, a visitor from Mars might reasonably suppose that the many constraints built into developmental and other evolutionary processes should stifle future variation. And yet—if you'll pardon yet another brief rhapsody—the kingdom of animals has produced a bewildering variety of innovations, including powered wings, camera eyes, arms with suckers and separate brains, external armor, electric organs, magnetic sensors, chemical lightbulbs, and a range of mating and childrearing systems that exceed the imaginations of even the most deranged or intoxicated science-fiction writers. Highly constrained developmental systems have managed to free rather than leash evolutionary innovation. The fact that this has happened begs explanation.

The remainder of this essay explores the many ways that developmental systems facilitate the production of adaptive variation and the underlying processes that may have granted them the ability to do so.

Developmental Bias Facilitates Evolution

The idea that biases in existing evolutionary processes can enhance future evolution is not new. In 1977, the Nobel Prize-winning French biologist François Jacob characterized the processes of evolutionary progression as "tinkering" or "bricolage" (art constructed of mixed media), more often recombining existing adaptations than inventing new ones.[8] Provided that those existing adaptations are sufficiently general in their uses, this method of construction can be more efficient than one that continually invents new adaptations. Fortunately, evolution has spontaneously discovered numerous general-purpose structures, tools and methods; if it hadn't, the dead hand of the past would indeed be quite constraining.

During the 1990s, deepening knowledge of embryonic developmental processes suggested additional specific mechanisms that could enhance evolvability. At around the same time, evolutionary biologist Mary Jane West-Eberhard contributed profound insights into the two-way relationship between within-life phenotypic adaptations and the longer timescale of genetic adaptation, culminating in her landmark 2003 book *Developmental Plasticity and Evolution*. In 2005, systems biologist Marc Kirschner of Harvard and molecular biologist John Gerhart of the University of California, Berkeley collected these developing ideas into a general theory of facilitated variation and presented it in the book *The Plausibility of Life*. (This excellent and accessible text is listed in the Major References section at the end of this essay, as is West-Eberhard's monumental and far more technical one.)

Somewhat rephrased and simplified for the purposes of this essay, the primary elements of the theory of facilitated variation can be summarized as follows:

- **Biology utilizes modules that can be connected and disconnected easily**. Biological construction processes are typically modular, dividing naturally into largely discrete units that interact with one another at only a limited number of points while being tightly ordered internally. Modules can activate sub-modules and sub-modules activate sub-sub-modules. The connections between modules are said to be weak because they can be changed easily, like unplugging a plug from one socket and plugging it into another. Importantly, these connections do not pass on instructions. When one module activates another, it merely issues the command, "Do that thing you do right here and now," perhaps accompanied by an intensity parameter. Weak linkage allows modules to be rearranged arbitrarily and permits bricolage.
- **Adaptation largely occurs through the recombination of traits**. Evolution seldom creates entirely new traits but rather produces its abundant novelties by mixing, matching, recombining and tweaking traits that already exist. Many of

these traits are highly general-purpose in their uses; below, I call them skeleton-key widgets.

- **Developmental systems use "smart" construction processes**. Many of the construction methods that operate in embryonic development are inherently flexible, possess a certain sort of algorithmic intelligence, and can adapt on the fly to changes in body structure. The capacity for local initiative possessed by these "exploratory" and "adaptive" processes reduces the number of mutations necessary to accomplish an adaptive modification.
- **Evolved adaptive responses guide further evolutionary adaptation**. The developmental and within-life responses of organisms to changing environmental conditions can channel directionless mutations in adaptive directions through a process often called the Baldwin effect.

Kirschner and Gerhart make a convincing case that these several mechanisms significantly enhance evolvability and help explain why animal (and plant) evolution has been so rapid and innovative. Their thinking is crystal-clear and their analysis lucid as they offer deep insights into how evolution does its work. However, they make no attempt to unite all these processes into a single paradigm. Furthermore, they freely admit that they do not know *how* evolution managed to produce exploratory processes, modularity and weak linkage.

Watson's theory of natural induction through self-modeling adds an additional element of facilitated variation and also offers a novel, persuasive explanation of the processes that drive facilitated variation into being. Reduced to its bare essentials, the idea goes as follows:

- **Developmental systems channel the effects of mutations in adaptive directions by generalizing on past evolutionary experiences.** Although Earth environments constantly change, they do not do so freely. Typical Earth conditions and the ways they change possess patterns, recurrent characteristics and many structural regularities. As organisms evolve adaptations to contend with changing environments, these regularities leave behind an imprint: Adaptations to conditions that typically shift together become developmentally linked, while those that track independently changing conditions become or remain independent. As a result, the genotype-phenotype mapping function takes on properties that track the correlational structure of the world. When a population of organisms face environmental challenges, this structure in effect generalizes and raises the likelihood that directionless mutations will yield adaptations responsive to those challenges.

The theory of natural induction lies at the cutting edge of evolutionary thinking, and it is neither completely worked out nor proven to be true. However, for many who have examined Watson's ideas, the insights and modes of analysis he offers ring the bell of deep explanation; at the very least, it seems that something *like* what Watson proposes must occur. By the end of this essay, perhaps you will agree.

But Watson's theory offers much more than intellectual inspiration. Research is inevitably guided by theory, and his paradigm suggests that researchers studying developmental systems and other aspects of gene regulation should keep their eyes open for certain types of findings that they might otherwise miss.

In the following pages, I explore Kirschner and Gerhart's ideas and their basis in the current understanding of developmental processes. My presentation differs from the original in that it applies Watson's general mode of thinking throughout. The last third of the essay focuses on Watson's ideas specifically.

Let's dive in.

Evolutionary Design Motifs

Humans assemble or construct structures; organisms grow them. Nonetheless, there are important similarities between human construction and biological growth processes. Both utilize broadly useful elements that can be flexibly combined. Houses are built from recurrent elements such as nails, two-by-fours, drywall, insulation and paint. Animal bodies, too, are built out of standardized parts, most prominently individual cells but also structural substances such as keratin (the primary constituent of nails), claws, antlers, hooves, hair, fur, scales, feathers and the surface layer of skin. Large-scale structural motifs abound as well, for example, ductwork, electrical conduits and structural beams in houses and blood vessels, nerves and bones in animals.

In one sense, construction limited to a finite set of standardized elements is a constraint. But if the elements are chosen well, they can be combined to build an infinite variety of final objects. The same power of combination is seen in chemical elements, the fundamental substances of which everything on Earth is composed, animate and inanimate alike.*

In addition to general-purpose structural constituents, humans and biology have also discovered the power of general-purpose *tools*. Hammers, wrenches and screwdrivers are found in every tool chest because they are commonly useful; evolution, too, has found tools that it uses repeatedly, such as continuously variable parameters that control the timing of developmental processes and hormones (and other signals) that diffuse away from their source and affect all cells within a given distance.

---

* In a deep sense, the flexibility of life depends on the flexibility of chemistry. But it goes deeper than that: Chemistry's numerous talents emerge from the widely varying properties of atoms that differ (almost) only in the number of protons they contain in their nuclei. The universe is composed of interchangeable parts all the way down.

Both human builders and biological systems have also found value in *modularity*, the technique of constructing complex structures and processes out of largely independent sub-structures and sub-processes. Modularity is a general-purpose design paradigm that facilitates the construction of novel structures.

We will discuss some of these design motifs in later sections. But first, let us examine the sophisticated control systems that create them: the gene regulatory networks embedded into the DNA of eukaryotic organisms

Eukaryotic Gene Regulatory Networks as Biological Computers

For several decades after its discovery, DNA was conceptualized as a library of genes: sequences of base pairs that are first transcribed into RNA and then translated into proteins via the genetic code. However, when the human genome was fully decoded in 2003, researchers were astonished to find that only about one to two percent of it consists of genes that code for proteins. For about five intellectual moments, the remainder was described as "junk DNA." We now know that much, and perhaps most, non-coding DNA consists of regulatory elements that control when and where genes are expressed.

The system that manages gene expression in eukaryotic organisms is a greatly expanded version of the on/off switches, operons and signaling proteins found in prokaryotic DNA. While prokaryotic switches are short sequences of DNA, perhaps five or six base pairs in length, their equivalents in eukaryotes extend for hundreds of base pairs. Each of these switches can take numerous signaling inputs and may act on multiple genes. As an additional wrinkle, DNA strands can bend, twist and squirm in such a way that more distant regulatory sequences on the same strand contact the primary switch and alter its operating characteristics.

Biologists have given specific names to these switches and their influencers, most prominently "promoter," "enhancer" and "silencer."

Collectively, they are called cis-regulatory elements, where "cis" essentially means "somewhere or other on the same DNA strand." From the perspective of a cybernetic analysis of gene regulation, all cis-regulatory elements impacting a single gene, together with a few other stray regulatory processes with similar functions, can be grouped together and discussed as a figurative "gene control box."

Gene control boxes are, by default, set to the OFF position, meaning that the gene they control remains quiescent, safely wrapped around a protein spool. However, certain incoming signals supply a "vote" for "flip to the ON position." Others vote to keep or turn the switch off.

The system is not democratic in that some votes count more than others. When signals with opposing messages reach a gene switch, the switch weights all the votes and then sums the weighted values. If the combined signal adds up to an ON command, the portion of the DNA double helix that codes for the gene unspools, and gene transcription begins; if the signal adds up to an OFF command, gene transcription is prevented or, if already in progress, halted.

DNA is a largely digital device; even though the mechanisms that manage it consist of chemicals, those chemicals have been domesticated into digital behavior. In bacteria, switches can only switch ON or OFF; there is no "half-on" condition. Smooth control of protein production is achieved by systems that regulate how long those switches remain on. However, in eukaryotic cells, multiple signals can combine in such a way that they not only activate a gene but also adjust the *rate* of that gene's expression. In effect, signals transmit a number that ranges from +1 to −1, where +1 produces maximal activation and −1 maximal suppression.

As described in the discussion of prokaryotic gene regulation, the signaling proteins that play a dominant role in initiating or terminating gene transcription are unimaginatively termed "transcription factors." These transcription factors are themselves encoded by genes, and each has its own control box that takes a range of inputs.* One of these inputs may be the same transcription factor that the control box activates; this permits a form of negative feedback to cap the production of the factor. However, many other inputs have a say in determining when and where that transcription factor is expressed, most especially the presence of yet other transcription factors. It commonly happens that when a control box is flipped to the ON position, it triggers the expression of a transcription factor that goes on to activate another transcription factor, yielding a cascade of signals that may reach through dozens or hundreds of downstream levels.

Taken together, these (and other) DNA-hosted processes form a complex, versatile and reliable control system called a gene regulatory network (GRN). The same term is used for the entire system and its various linked and nested subsystems.

The elements of GRNs can function as familiar logic circuits. If gene A is activated only when transcription factors B and C are present, this arrangement constitutes an AND gate. If either B or C can switch on A, we have an OR gate. (However, many and perhaps most GRN internal relationships are non-Boolean.*)

Some GRN circuits may behave as oscillators.[10] For example, activation of gene A may, in turn, activate gene B and turn itself off; then, activation of gene B activates gene A and turns gene B off. Since the production of proteins takes time, such oscillators can function as clocks. Some GRNs may possess two or more distinct stable states, switching from one to the other under the influence of signaling inputs.

Regulatory systems link genes in a complex web of interactions, causing some to be expressed together and others separately at various times and in various ways. Any given gene may be included in multiple distinct collections of genes depending on which GRN is currently

---

* Because the genes that encode transcription factors often reside on different chromosomes than the control boxes they affect, they are called trans-regulatory elements.

running the show. In this respect, they resemble humans, who maintain simultaneous membership in families, occupational organizations, nations, religions, classes and political movements.

GRN operation is already complex even in unicellular eukaryotes, but it becomes more complex still when multiple cells band together in colonies. In such crowded conditions, protein products and bioelectric signals produced by cells located near one another form a web of environmental inputs that can produce emergent, higher-level effects. A simple process of this kind known as "quorum-sensing" occurs even in prokaryotes. When population densities in a colony of bacteria or unicellular eukaryotes reach a certain threshold level, intercellular signaling triggers a rapid, collective change in gene expression patterns in members of the colony, altering and presumably optimizing their behavior and physical characteristics.

Permanent multicellular organisms take this complexity up another notch. Local signals are augmented by medium- to long-range signals produced elsewhere in the body, such as endocrine hormones and nerve impulses under the control of a central nervous system. Furthermore, organisms that possess symbiotic partners—which may include all animals and plants—often evolve to interpret chemicals produced by their symbiotes as signals and rely on them to trigger developmental and within-life processes.[11] For example, the intestinal tracts of mice cannot develop properly except in the presence of the symbiotic bacteria that colonize mouse intestines.[12]

When a hormone or other signal arrives at a cell membrane, it triggers the release of an internal signaling protein stored at the cell's membrane for just that purpose. That internal signal may trigger other internal signals that ping-pong around the cell until a signal reaches and binds to a control box on one of the GRNs puppeteering the DNA protein-synthesis machine. GRNs can also be locked into alternative states by proteins that bind strongly to a control box and force it into the ON or OFF position. Such semi-permanent switch-locking is one of the primary means by which the various kinds of cells in animal bodies take on their radically different forms despite possessing identical DNA.

Thus, contrary to typical imagery, DNA is not a static molecule about which transcription/translation systems hover like bees, placidly reading genes, deciphering the genetic code and building proteins. Rather, DNA is an active device in its own right, forever spooling and unspooling, folding and unfolding, clicking and clacking away as directed by its internal control systems as they operate independently and also respond to incoming signals from other cells and the external environment. Embryonic development of an adult plant or animal operates through the well-timed switching-on and -off of gene transcription as directed by this system of intra- and intercellular computation, and the GRN concept is most well developed with respect to developmental processes. However, GRNs also manage the hour-to-hour function of the individual cells of a fully formed organism. The total collection of GRNs in all the cells of a multicellular organism functions as a kind of biological computer composed of biological computers, and it acts to build and maintain the organism throughout its life cycle.

Of course, because this is biology we are talking about, these systems are not designed in the linear fashion favored by humans. GRNs contain recognizable logic elements and recognizably perform computations, but they also utilize dynamic relationships that are difficult (although perhaps not impossible) to analyze as static computational circuits.[13] In addition, they include numerous redundant connections with seemingly near-identical functions. Highly refined feedback processes are backstopped by others that operate more grossly but can get the job done in a pinch. GRNs are computers, but they do not resemble computer systems that humans have ever built.

Still, there is much to be said for the computational architecture favored by life. Among other useful characteristics, gene regulatory systems are robust and much less susceptible to locking up than human-built computer programs. This resilience is an essential, necessary quality for many reasons, not the least being that developing embryos cannot be rebooted if they were to crash.

GRNs also need to be able to tolerate high levels of noise. Although enzymes control chemical reactions remarkably well, they cannot do so perfectly, and unintended chemical products will inevitably appear on the scene. In addition, thermal energy causes every molecule in every cell to shimmy, rotate, bounce and vibrate. These unpredictable influences can interfere with all biological systems, including GRNs. Nonetheless, due to the presence of redundant, compensatory feedback systems, embryonic development and daily housekeeping functions almost always work as intended. So powerful is this redundancy that researchers can greatly disturb a developing embryo, even to the point of disconnecting body parts and re-attaching them elsewhere, and it will often be able to restore itself to the original plan.

And yet, despite their robust stability, GRNs are also subtly attuned and responsive. Certain specified inputs can cause organism function to shift markedly, perhaps dramatically. For example, in the presence of elevated temperatures, GRNs can suddenly begin to express genes that code for "heat shock factors:" special proteins whose only purpose is to resist heat stress.

The structures of GRN circuits are so intricate and their dynamic interactions so subtle that it currently takes research laboratories many years to fully work them out even in part. Some circuits are constructed in such a way that they can control elements of embryonic development in a smoothly varying manner; these are designed to be easily shifted by mutations and act like dials or sliders. Total body size is a typical example of such a continuous parameter, as exemplified by the variation seen among breeds of dogs and the frequently observed evolution toward decreased size that often occurs in large mammals such as elephants when they become trapped on islands. Other circuit designs can produce discontinuous changes in body structure through the manipulation of GRN elements that act as switches. These switches can be thrown either by incoming signals or by mutations. The full power of GRN switches can be seen to operate when an organism expresses an atavism (the sudden return of a lost ancestral trait), for example, the teeth that occasionally appear in chickens.

The total set of GRNs of an organism constitute its genotype-phenotype mapping function.

Hans Spemann, the first scientist to properly explore embryonic development, was so overwhelmed by its intricacy that he sometimes believed himself to be observing the imprint of psychic forces on matter.[14] He had difficulty conceiving a materialistic explanation of what he was seeing because nothing like a computer had yet been invented.

GRN Outputs: Proteins

Computer programs receive inputs and produce outputs. GRNs take incoming signals as their inputs and produce outputs in the form of proteins. The GRNs that control embryonic development do so by producing the right proteins in the right places at the right times.

Human DNA yields perhaps 100,000 proteins created through the splicing, dicing and variant readings of about 20,000–25,000 underlying genes. Many proteins function as effective and specific chemical catalysts called enzymes, molecules that speed up certain chemical reactions by many orders of magnitude while leaving others entirely alone. Given the critical role that enzymes play in metabolism and DNA transcription—the most basic functions of life—enzymes must have emerged early in life's origins. One enzyme of particularly ancient lineage is ATPase, a rotating protein complex that allows organisms to extract energy from life's "universal energy currency," the molecule ATP.

Proteins that have evolved more recently include those that correct genetic code transcription and translation errors, splice and dice protein products into new proteins, wrap DNA around chemical spools, carry signals, guide the folding of other proteins and, in general, perform all the housekeeping functions required by modern organisms. Other proteins, the transcription factors mentioned above, participate in the control of gene expression. On a more observable scale, structural proteins form skin, feathers, horns, beaks, hair, muscles, tendons, ligaments, bones and teeth.

Leaving aside those whose only use is to send signals, proteins can be said to form the raw materials of life, its building blocks and its molecular machinery. These materials vary relatively slowly over evolutionary time, and for this reason, organisms with distant common ancestors tend to share many genes (or, at least, very similar versions of those genes). An organism's initial construction and ongoing operation require supplies of these raw materials in specified amounts and at specified times, and GRNs resemble the supervising contractors/engineers that manage the process. Any mutation that alters the structure of a GRN has the potential to alter the form and function of the organism that emerges, in essence providing the supervising contractor with an altered design plan.

But if genes are relatively static, and GRNs do most of the "creative" work by controlling how genes are used, this suggests that many of the mutations that make a difference affect GRNs rather than literal genes. They are not "genetic" mutations but mutations in non-coding sequences, the regulatory structures that I have grouped together as gene control boxes. The net effect of such mutations is to deploy the same genes differently.

But most of us have become used to thinking about mutations as alterations to genes. Let us take a moment to further disassemble that outdated idea.

Mutations Reprogram GRNs

Protein-coding genes occupy no more than one or two percent of human DNA, while cis-regulatory sequences—gene control boxes—may occupy as much as 50 percent of the total (although the actual figure is contested due to various methodological issues).[15] If the true percentage is even 1/20th as great as this estimate, mere probability suggests that mutations should occur more frequently in regulatory sequences than in genes. Regulatory mutations in effect rewrite GRN software rather than altering the genes that the software controls.

Even before the Human Genome Project unexpectedly revealed that only a small percentage of DNA codes for proteins, biologists working in the field of embryonic development had begun to suspect that most significant evolutionary innovations arise from mutations in regulatory sequences rather than in genes. This idea was raised as early as 1969 by Roy Britten and Eric Davidson and was subsequently advocated with enthusiasm by Mary-Claire King and Allan Wilson in the 1970s.[16] In the 1990s, researchers in the field of evolutionary developmental biology pressed the same idea again.[17]

These prescient scientists had reached this conclusion through logical reasoning that went something like this: Many features of organisms are constructed of the same proteins and hence produced through the activation of the same genes. Mutations in those genes should, therefore, affect numerous features all at once, an effect called pleiotropy. However, it is an observed fact that numerous features of organisms evolve independently; for example, there is no linkage between the length of a dog's snout and the length of its tail, even though many of the same genes are used in both. This independence suggests that the evolution of body form depends more on changes in how and where genes are activated than on changes in the genes themselves.

The problem of pleiotropic effects does not arise for genes used only for a single function or in a limited anatomic region; these are called tissue-specific genes. For example, the proteins that respond to light within the rods and cones of the retina appear nowhere else in the body and can, at least in principle, vary freely without risk of pleiotropic consequences. However, if shifts were to occur in the genes that build cartilage, bone, skin, nerves or muscles, they would reverberate in too many places to be useful.

Furthermore, it would seem logical to assume that the genes used for ubiquitous "housekeeping" functions, the day-to-day needs of an operating cell, have become so optimized over evolutionary time that most variations would either do nothing or cause harm. Subsequent comparative genomic analysis has supported this theoretical reasoning and shown that housekeeping genes do indeed change slowly over evolutionary time.[18] The major exception occurs during periods of major evolutionary changes, such as the shift from unicellular to multicellular life. Animals simply have housekeeping needs somewhat different from bacteria.

Similar theoretical reasoning suggests that the genes that code for transcription factors and other signaling proteins should also change infrequently. Structural proteins must physically build structure and enzymes earn their keep by catalyzing real reactions, but transcription factors are just arbitrary codewords; they attach to gene on/off switches that are constructed in such a way as to receive them. If a transcription factor were to change, all its many connections would break. Because of this, it's hard to think of a situation where altering a signaling protein would be a good idea; Morse code has remained unchanged for much the same reason. However, some signaling proteins are tissue-specific, and alterations to them might plausibly produce coherent effects.

Hard evidence has subsequently supported this logical inference, too. While transcription factors can occasionally change in meaningful ways, most seem to be terrifically stable. For example, Pax-6, a transcription factor that plays a major role in the development of visual systems, remains functional in fruit flies when the mouse version is artificially substituted for the fruit fly version.[19] Mouse and fruit fly Pax-6 proteins are not identical down to every amino acid, but they are sufficiently close that they function identically despite the more than one billion total years of evolution that separate the two organisms. Just as remarkably, the structure of mouse Pax-6 is precisely identical to its human equivalent, down to the last amino acid.[20] Since our common ancestor with mice lived about 100 million years ago, this indicates complete sequence preservation over a total of 200 million years of independent and considerable evolution.

Based on these theoretical arguments and various forms of confirmatory evidence, many evolutionary biologists have by now come to believe that meaningful variations result frequently or even usually from mutations in gene control boxes—the regulatory elements that activate or halt gene expression.[21]

Some mutations in control boxes alter the "voting power" given to a particular input or reverse its sign (switching its meaning from "ON" to "OFF," or vice versa). Control boxes can also mutate so that they change their "opinion" on what signals they are "interested" in receiving at all; mutations of this kind rewrite the topology (connection structure) of GRN circuitry. At the level of final output, changes to control boxes may have no effect at all, modify a continuously variable parameter such as total size or produce a discontinuous, sudden change in form and function. If a control box stands at a critical juncture, alterations to it may alter phenotype dramatically, even going so far as to add a new, macroscopic body part (for example, chicken teeth).

Mutations in transcription factors can also produce useful variations by disconnecting one or more control boxes from modules to which they were previously linked.

The realization that mutations are not necessarily literally genetic has been slow to penetrate. Medical researchers are always said to be looking for the "genes" behind autism, schizophrenia or Parkinson's disease, but regulatory mutations have also been shown to play a role in many inherited medical conditions.[22]

It does make particular sense that defective genes should cause pathological states. However, for the reasons given above, mutations in genes are far less likely to produce *useful, adaptive* variations. It is, therefore, somewhat confusing to laypeople that biologists continue to use the term "genotype" to describe the DNA underpinning of an organism's inheritance. To many (but, surprisingly, not all) biologists, "genetic" and "genotype" now simply refer to the totality of DNA sequences, whether non-coding or coding. However, the terminology remains misleading for non-specialists. I've taken an informal poll of three physicians of my acquaintance, two medical researchers and one PhD botanist, and all six defined "genotype" as "the set of all protein-coding genes."

The mistaken idea that evolution is necessarily driven by gene mutations has been so strongly driven into many people's minds that it is perhaps worth providing a few analogies to weaken that preconception.

- The originality of newly written books does not lie in their use of new words or new letters.
- New recipes seldom use newly-discovered foods.
- All houses are composed of much the same materials, for example, nails, shingles, two-by-fours, drywall, vents, blowers and copper wire; the main difference between houses is how those materials have been put together. Materials do change occasionally; the ranch-style homes commonly built in the United States today are constructed from substantially different materials than thatch huts. But, over shorter periods, new and novel forms of home construction rarely involve new and novel materials. Most home variation emerges from modifications to home building *plans*.

An emerging standard employs the term GRN to refer both to the various regulatory systems that control gene expression and the genes they regulate, and I will usually follow that practice in what follows. However, in some places, especially when quoting papers, I will write "genome," "genotype" or "genetic" when I mean "genes plus GRNs" because all these expressions are becoming synonymous.

Let us step back for a moment and note something remarkable: Over the course of evolution, animals and other organisms have evolved biological computers that can be usefully reprogrammed by the action of random mutations. If you randomly modify lines of code in a typical program made by humans, it will probably crash. Evolution has produced a special kind of computer that can tolerate random reprogramming and, in many cases, make profitable use of it. In GRNs today, most mutations have no obvious effect, relatively few cause organism failure, and an elevated percentage produce coherent, functional and perhaps even specifically adaptive outputs (Uller et al. 2018, p. 955-956). It's quite an accomplishment.

Modularity

After this long introduction, we can at last begin to discuss the various elements of Kirschner and Gerhart's theory of facilitated variation, beginning with modular design.

Modularity is a universal characteristic of organisms, operating at many levels of their organization.[23] As the term is used in evolutionary biology, it means that organisms and their actions tend to divide naturally into largely discrete units that interact with one another at only a limited number of points while being tightly ordered internally.

The cleavages that separate biological modules operate in space, function and time. The cells of a multicellular organism are paradigmatic evolutionary modules, but the same pattern is pervasive. The distinct regulatory and active sites on proteins can be regarded as intramolecular modules, as can bacterial operons. Within eukaryotic cells, biochemical processes are frequently confined to internal cellular compartments so that they do not interfere with one another; for example, energy production occurs within bounded mitochondria, photosynthesis within chloroplasts and DNA transcription inside the nuclear membrane. Organs are bounded collections of tissues with deep internal integration and a limited number of specific interactions with other organs.

Much the same is true of modularity over time: When organisms develop through multiple life stages, such as caterpillar/butterfly or tadpole/frog, each stage is sufficiently autonomous that it can undergo some evolutionary changes without much impacting the characteristics of the other stages.

Modularity can also be seen, if more subtly, in homologies, structures that gradually morph into different forms and take on new functions over evolutionary time while retaining a clear, demarcated, continuous identity.[24] For example, the bones of the mammalian inner ear can be traced back to their origins in the jaw bones of ancient reptilians.

Modules are commonly arranged hierarchically. Top-level features such as limbs are composed of sub-features such as the individual bones, joints and muscles that compose them, and each of those sub-features is built of sub-sub-features such as fibers and protein complexes. Low-level features tend to be used as standardized parts in the construction of many higher-level ones.

Fabrication out of discrete modules offers numerous benefits. Many of these relate to flexible design; children (and adults) can easily build a vast range of structures out of just a few kinds of Lego blocks. When mutations in GRNs lengthen a limb, move a forward-facing eye to the side or increase the density of hair follicles, they are doing the equivalent of rearranging Lego blocks or adding more of the same kind of pieces. Modular design also enhances stability. A fully interlinked structure may break entirely if a single part fails, but structures composed of separate modules can confine the damage to a single region or function. Both of these effects raise the chances that GRN-altering mutations will produce useful results.

In a pedagogically ideal world, each modular feature or trait of an organism would be constructed by a discrete GRN module and each sub-feature by a sub-module. Alas, it's not that simple. GRN circuits do possess significant modularity,[25] but the correspondence between circuits and features is (probably) only sometimes crisp.

Limb formation is one of the best examples of straightforward correspondence between independent GRN modules and independent modular traits. When certain signaling proteins called growth factors are applied anywhere within a range of positions in a chicken embryo, a fully developed leg grows from the site; different growth factors cause a wing to grow there instead.[26] Limbs and wings are physical modules that are constructed by distinct, self-contained, limb- or wing-constructing GRN modules. When the appropriate growth factor reaches the right control box, these GRNs wake up and get to work. Once a top-level limb-building module is underway, it (more or less) goes on to activate sub-modules that construct sub-features such as joints and fingers; in turn, the system for building fingers invokes others that build fingernails, and so on. One might regard all these subsidiary structures as mere components of a single limb-construction module were it not that multiple distinct higher-level systems can utilize the same sub-modules; for example, jaws and limbs have quite different evolutionary histories, but they both include bony joints.

However, in many other cases, the correspondence between GRN modules and modular traits is not at all clear-cut. Some GRN modules may be functionally but not structurally integrated in the sense that they jointly contribute to a final effect. For example, tongues, lips, vocal cords and many neurological elements together form the behavioral module we call human speech, but there is no single "speech system-creating module." In addition, embryonic development is wildly iterative; one process may affect another that then goes on to alter the first, and the modularity of the result can't necessarily be traced to a modular cause. Evolution follows biological design principles, not those used by humans, and, as such, their mechanism and structure need not strike a human observer as plainly obvious.

We lack good language to describe deeply iterative processes like embryonic development—if, indeed, there is anything quite like it—and our standard intuitions don't model it well. To keep the following discussion understandable, I will sometimes speak of "GRN modules" as if they correspond to modular features, with the understanding that the expression is imprecise. I will also conflate GRN modules and sub-modules with adaptations and sub-adaptations, features and sub-features, traits and sub-traits and processes and subsidiary processes. None of these correspondences should be taken too literally, and I will periodically issue a reminder to that effect.

Perhaps it is worthwhile taking a moment to point out that, except in certain simple, well-demarcated cases, even such terms as "traits" and "adaptations" are simplified, idealized abstractions that do not have exact counterparts in the far messier world of biology. How many traits does a hand possess? Which parts of the mouth are adaptations for eating, and which are for speech? When we try to discuss complex systems, we can seldom avoid using intuitively meaningful expressions that are difficult to define. For example, in medicine, we refer to disease entities as if they were well-defined categories, even though no two people with COVID-19, lupus, asthma or schizophrenia exhibit exactly the same characteristics or respond to treatment in the same way. Similarly, in politics, we speak of demographics such as "evangelical," "working-class," or "college-educated" as if they were distinct human species although they are not even discrete groups. It's even difficult to specify in any universally applicable way what we mean by "species."* The use of such shorthand expressions, while necessary and unavoidable, risks intellectual self-deception. As famed evolutionary biologist Richard Lewontin wrote (quoting others), "The price of metaphor is eternal vigilance" (West-Eberhard 2003, p. 81). Keeping this in mind, let's return to modularity.

One of the great advantages of biological modularity is that it permits distinct structures to evolve independently; kidneys can come under selection for new capacities without altering the structure of livers, even though both are constructed using many of the same genes. Mutations in the control box that activates a GRN will affect the feature that the GRN produces and no other (or, at least, not directly).

However, some traits are linked and cannot easily undergo independent evolution. For example, quadrupedal animals possess legs that are entirely symmetric between right and left and mostly symmetric between front and back. Any mutation that alters one of the legs will alter all the others. This fixed relationship suggests that

---

* See also the discussion of "objectness" in the essay in this collection titled "The Nonlinear World of Yoshitsugo Oono."

the process of constructing front and back legs employs one or more shared GRNs.

The patterns of linked and independent biological modules reflect and, in a certain sense, model Earth environments. Some environmental features of the world vary together while others vary separately, and the relationships between the modules that respond to them track this correlational structure. The modular characteristics of organisms and the relationships between them are thus generalizations about the variational properties of the world. (Much more on this important idea below.)

Weak Linkage or Pluggable Modularity

The example of chicken wings growing in the wrong part of an embryo when growth factors are applied there demonstrates an important characteristic of how GRNs function: When an incoming signal activates a GRN module, it doesn't pass along detailed instructions but merely transmits an activation command. The use of simple, low-information signals permits GRN modules to be freely connected and disconnected.

Suppose that when control box A is activated, it produces a signal S that binds to and activates the control boxes of sub-modules B, C and D. Suppose, next, that environmental conditions change so that in cases where sub-modules B and C have been activated, the structures or processes produced by sub-module D become problematic and reduce fitness. If in any organism, a single mutation occurs in the control box for D that causes it to lose affinity for signal S, D will be delinked from B and C. Since, according to the terms of this scenario, D reduces fitness in the new environment, organisms possessing this mutation will come under selection and soon dominate. The same process can operate in reverse, appending new sub-modules to an existing set.

Given the ease of this process, one may say that the linkages between modules aren't soldered into place but consist of plugs that can be inserted and removed. Kirschner and Gerhart describe this characteristic as "weak linkage,"[27] but perhaps "pluggable modularity" is a better, less jargony term.

Humans use analogous methods when they design and construct homes. An architect can specify "build a non-load-bearing wall here" without appending instructions about the best method of affixing drywall. They merely indicate the location of the wall and its size and rely on established practices to provide the necessary detail. This lack of necessity for information transfer continues down to deeper levels. People who hang drywall use nails but need not teach nail makers how to manufacture, package and ship them, and nail makers don't teach iron miners how to build iron-mining machines. The modular depth of an ordinary home is substantial if one considers the origins of its individual parts and the origins of the parts that compose or construct those parts, but architects can take many of those subsidiary details for granted and issue relatively schematic construction diagrams that address top- and middle-level requirements. This greatly simplifies the task of designing buildings.

Pluggable modularity provides evolution with a similar structured-but-flexible design capacity. Suppose that the ancestors of tyrannosaurs were quadrupedal, but circumstances arose in which having shorter front legs than back legs offered a fitness advantage; in that case, mutations that partially severed the linkage between front and back legs would be favored under selection. Similarly, if in a beetle species without horns, mutants with horns had an advantage, any mutation that activates a leg-building module in the "inappropriate" location of a beetle's head would be selected for and could later be tweaked into optimal shape. If continuing a developmental process for a longer or shorter time offers a fitness advantage, a mutation that appropriately modifies the timer would similarly come under positive selection.

While analogy with human activity makes it easier to understand how pluggable modularity enhances future evolution, it doesn't help explain its biological origin. Humans use modules because we have learned through trial and error that this is the most efficient and cost-effective way to design and build new structures and devices, but evolution doesn't work that way. Selection can only choose adaptations based on their *current* fitness, and modularity doesn't by itself enhance fitness. Non-modular hacks often work better for any given problem, which is why computer programmers are constantly caught using them, no matter how severely they are instructed not to. The fact that pluggable modularity might aid future evolution is of no relevance to short-sighted selective processes. Given this, one has to wonder why biological organization is so thoroughly modular.

As I will describe in this essay's epilogue, Kirschner and Gerhart make a half-hearted attempt to ground the origin of pluggable modularity in a form of meta-selection—selection on lineages—but it doesn't work very well. Watson's theory has attracted attention in part because pluggable modularity emerges quite naturally from it. To summarize in one sentence what I will later explain in detail, there exist numerous regularities in Earth conditions and how they change, and modular design reflects the imprint of those regularities on developmental processes.

No matter how it arose and is maintained, pluggable modularity facilitates evolutionary variation. To illustrate its power, let us examine a famous example.

Modular Eyes

In the 19th and 20th centuries, the presence of similar structures in evolutionarily distant animals was said to illustrate "parallel" or "convergent" evolution—the independent discovery and refinement of structures to accomplish similar purposes. This description remains correct when applied to the evolution of flight in bats and birds; although both have wings, they evolved their wings out of different body parts and use entirely different GRN modules to build them. Bat wings and bird wings are analogous rather than homologous; they are similar in form and function but not in origin.

However, many other forms of apparent analogy have recently come to be understood as examples of hidden homology; similar structures in distantly related organisms are often built in much the same way, using many of the same modules. One oft-cited example is the surprising similarity between the embryonic development of eyes in mammals and octopuses.

Octopuses, like humans, possess "camera eyes," structures that pair a focusable lens with a light-sensitive retina. Given that the most recent common ancestor of humans and octopuses was a minute flatworm whose light sensors consisted of simple eye spots, the striking similarity of human and octopus eyes seemed such a clear case of convergent evolution that it was often used to illustrate the concept. However, when researchers acquired the ability to study developmental processes in detail, they discovered that the same transcription factor, Pax-6, initiates eye development in both octopuses and humans. Moreover, at least 70 percent of the same genes are expressed during the two processes.[28] Some of these are undoubtedly universal housekeeping genes, but most are not; the similarity of gene expression between octopus and human eyes is much greater than that between octopus eyes and, say, human connective tissue. Considering that our common ancestor with octopuses lived at least 500 million years ago, this overlap is remarkable.

The ancestral flatworm apparently possessed light-sensitive cells overlaid by a protective layer of transparent tissue. The embryonic development of this ancient flatworm's eyes seems likely to have been initiated by Pax-6, given that Pax-6 still initiates eye development in both the invertebrate and vertebrate descendants of that ancient creature. Camera-eye evolution would have begun when organisms came to depend sufficiently on vision that selective pressures arose to improve it. The stages of what followed remain obscure but may have resembled the following:

First, the structure of the protective transparent tissue began to alter, in effect adding the task of "focusing light" to its initial job description as a protective film. Nearby objects can be brought into partial focus even when a lens lies almost directly on top of a light receptor, but to resolve objects further away, the laws of optics require a greater distance between the lens and receptor. As the value of vision grew, the emergence of structures that could increase the distance between primordial lenses and primordial retinas came under selection. Animals possess a generic ability to create liquid-filled spheres, which they use to build abscesses, cysts and friction-reducing bursae. Switch on the control box that activates a cyst-creating module between a retina and a lens, and you have a camera eye. In this (entirely hypothetical) explanation of the origin of the liquid-filled eye, the modules activated by Pax-6 evolved to produce signals that bind to a control box that activates a cyst-forming module on site.

Once organisms came to depend on camera eyes, selection pressure emerged for an ability to shift precise focus between near and far objects as well as to look in multiple directions. Structures that push and pull are ancient and ubiquitous; they operate as individual filaments inside cells. But when used to move or deform larger structures, they are bundled together into muscles. To move an eye up and down and sideways, muscles need to be placed in at least four locations; to focus a lens, muscles must either deform it (the method used in mammalian eyes) or change its distance from the retina (as in octopus eyes). In time, the modules activated by Pax-6 came to produce the correct signals in the correct locations to activate the preexisting modules that initiate muscle formation. The resulting camera eyes in vertebrates and invertebrates are not identical, but neither are they fully independent; they converged on similar structures along similar paths.

Researchers in the field of embryonic development have come to recognize this as a general pattern: Once evolution has found an effective way to do something, it typically reuses it. That it does so makes obvious sense because it is always easier to repurpose a useful tool you already have than to invent something entirely novel. The master plan that builds animals has discovered good tools and good ways to combine them into more complex tools, and it "remembers" both the tools and the ways of combining them.

Humans, too, use tools to build tools and (most often) manufacture complex objects by combining more familiar, simple ones. Even our most radical inventions, such as silicon chips with billions of transistors in them, are constructed by machines made by machines, built up from an evolved pyramid of inventions from the simplest to the most complex.

But there is one major difference: We humans assemble our creations but animal GRNs *grow* them.

Growing a Multicellular Organism

For the multicellular creatures we call animals to emerge from the much more loosely organized colonies of unicellular organisms that came before, collections of cells had to master the art of creating a single cell that could represent the whole: an embryo. This accomplishment produced a major evolutionary transition described as a shift in the level of individuality.* Selection on heritable variation in animals operates (mostly) on the entire organism rather than the individual cells that compose it.

Adult animals don't just pop into being but grow from single-celled embryos through the complex process of embryonic development. From this basic fact, it follows logically that most evolution must occur through modifications to developmental programs. This fundamental aspect of animal and plant evolution was ignored by evolutionary biologists throughout most of the 20th century, largely because they couldn't do anything with it. Embryology was

---

* This description comes from *The Major Transitions in Evolution*, the mind-blowing, extremely influential and well-worth-tackling 1995 book by John Maynard Smith and Eörs Szathmáry (Oxford University Press, reprinted 1998).

so mysterious that it was treated as a black box, a mere machine for making adults.

But this situation has now changed. After long neglect and following many novel discoveries, embryonic development has finally taken its proper place as the acknowledged leading edge of animal (and plant) evolution. The field of science that studies the relationship between evolution and embryonic development is evolutionary developmental biology, affectionately known as "evo-devo." Although much more remains unknown than known, evo-devo is an incredibly active field. (There is no better popular science introduction to this subject than Carroll 2006.)

The fully developed bodies of complex animals contain hundreds or thousands of distinct cell types that originate from a single fertilized ovum.† These genetically identical body cells become anatomically and physiologically different from one another through the differential activation of GRN subsets of their shared DNA. The initial transformation of embryonic cells into their adult form is triggered by incoming signals. At first, they retain their ability to revert to their initial protean form, but at a later stage in cell differentiation, the GRNs of these transformed cells become locked into a permanent setting. This locking-in is accomplished in part by substances that irreversibly attach to portions of DNA and, in effect, "glue" control boxes into the ON or OFF position. Fully differentiated cell types differ in structure, function and response to incoming signals.

The evolution of animal bodies required the emergence of reliable processes to alter cell GRN dynamic states in certain ways in certain locations; skin cells are meant to appear on the outside of organisms, not inside their kidneys. Embryos manage this complex process through a glorious process of self-interaction.

---

† Estimates of the total number keep increasing. Recent analyses have found 3,000 distinct cell types in the human brain alone. Genetic Engineering and Biotechnology News. *Largest Human Brain Cell Atlas to Date Reveals Unprecedented Detail*. October, 2023.

In bilaterally symmetric animals, the initial embryo is patterned asymmetrically along the two major axes of animal forms: anteroposterior (brain to tail/mouth to anus) and dorsal-ventral (spinal cord to belly). As the cells of an embryo divide and develop, they produce signals that diffuse from where they were formed but become progressively weaker with distance. These chemical gradients "inform" cells where they are located in the developing embryo and guide their further behavior. Other signals act more locally. These *paracrine factors* alter the behavior of GRNs in nearby cells, causing them to undergo changes. These newly modified cells go on to produce their own paracrine factors that, in turn, alter the structure and function of the cells adjacent to them, often including the very cells that initiated their own changes. This recursive characteristic of embryonic development complicates attempts to link individual GRN modules to specific outputs.

Cells in a developing embryo can identify whether they are located at the surface layer, in an intermediate location, or closer to the center. This identification triggers the differentiation of cells according to their relative positions. Some signals received cause cells to undertake coordinated shifts in position. Once in their new places, new or old paracrine factors again go to work. Thus, embryonic development consists largely of a series of signals that trigger actions, leading to further signals that trigger further actions.

Humans do not build houses this way. Contractors refer to detailed drawings and construct houses upward from the foundation to the roof in a sequential, linear fashion. In embryos, something functionally resembling construction drawings must be contained implicitly in those programmed serial interactions. The method seems absurdly fluid, but embryonic development can be counted on to construct the intended organism in almost all cases. Evolution has mastered deep iterative processes that can reliably yield the desired results.

Current knowledge is insufficient to explain how organisms perform these reliable iterations, and because it is so unlike human construction techniques or features of the non-living physical world, we struggle to find useful intuitions, models or even the terminology to discuss how it works. All we can currently say is that modules turn on and off, and they are designed in such a way that their turning on and off leads to a definite output. In a sense, they resemble computer programs, which work by iteration, too. However, with the partial exception of deep learning systems, even the most complex computer programs are more straightforwardly organized than embryonic processes.

But embryonic development is doing something right, for it is exceedingly robust; the process yields a reliable output despite the presence of many disturbances.

Attractors

In mathematical terms, proper embryonic development is an "attractor" of the embryonic developmental system. Attractors are states of dynamic systems that will eventually be reached from a broad set of initial conditions. Think of a pendulum in a grandfather clock: If you randomly shove the pendulum or start it from any position, it may jerk around a bit but will eventually settle into the single behavior of moving back and forth at a constant speed determined only by the length of the pendulum and its weight. If you push or yank on it too forcefully, you'll break the mechanism, but within broad limits, no matter what you do, it will eventually revert to the same predictable pendular motion. That regular, periodic motion in a single plane is the pendulum's attractor.

Alternatively, attractors can be visualized as the final common destination of raindrops that fall in watersheds—regions of terrain that drain to particular rivers or bodies of water. If a droplet falls anywhere within a watershed, it will find its way by surface flow or through the ground to that watershed's single output. Another term for a watershed is "basin." When physicists discuss the attractors of any dynamic system, they use the term "basin" to refer to the range of inputs that eventually converge on a given attractor. Within the basin that extends from the Allegheny Mountains in the east to the Rocky Mountains in the west, (almost) every drop of water that falls as rain and does not evaporate will eventually reach the Mississippi River, pass through New Orleans and enter the Gulf of Mexico.

Just as the range of locations on which raindrops can land in a watershed for a particular outflow point constitutes the "basin" of the watershed system, the range of initial pendulum configurations that result in typical pendular motion is the basin of a pendulum system's attractors.

Dynamic systems can have multiple attractors. If we consider the distribution of rainfall over North America as a single system, these attractors are the rivers that emerge from distinct watersheds. A drop of water that falls on the east side of the continental divide ends up in the Mississippi, and one that falls on the west side eventually reaches the Colorado River.

A robust system has deep, broad basins around all its attractors. Because of this, it's easy to get an expected result and difficult to get an unexpected one. A raindrop that lands in Kansas will never end up in the Colorado River (unless, perhaps, a truck takes it there). To say the same thing differently, if you were a giant and you wanted to toss a water balloon in such a way that its contents end up in the Mississippi, you wouldn't have to aim very carefully. Attractors with basins as big as the Mississippi watershed are hard to miss.

Embryonic developmental systems also utilize attractors with big basins. A lot can go wrong, or at least go differently, and an embryo will still develop into a reasonable facsimile of its parents; a frog egg will produce a tadpole even if the egg is frequently jostled or the conditions under which it develops cause one side to be significantly warmer than the other.

It is possible to break the mechanism of a grandfather clock by perturbing its pendulum too violently. Similarly, the basin of conditions that produce a viable biological output does have limits—for example, if an egg gets too hot, too cold, or is subjected to sufficiently intense radiation, it will not develop properly—but within those limits, environmental variation will not harm what emerges. This remains true even if the egg is subjected to significant and sometimes grisly alterations. For example, scientists can scoop out a large number of cells from an embryo, and it will still form a normal organism. Or, if embryos are starved so that half their cells die, they usually still develop normally. And if a limb bud is transposed to the wrong part of a body, it may shift back to the correct position or transform into something that corresponds to its new location.[29]

The basin of attraction for a healthy, normal embryo is very large indeed. That so many things can go wrong and an embryo will still emerge properly might make one suspect that it possesses a hidden blueprint specifying its final form. But, according to almost everyone who has studied the subject, the template is implicit rather than explicit; it is stored within the whole of the iterative processes used by the GRNs that control development rather than in a particular location. It is worth taking a moment to reflect on this, if only because it is so difficult to imagine how such a system operates. Evolution has spent hundreds of millions of years exploring and mastering a strategy for engineering complex structures that we humans have never tried and know almost nothing about.

However it is that those iterative systems do their work, they do it well. Embryonic developmental systems are inherently resistant to noisy influences of all kinds; you can't easily throw them off stride. It is as if they *relax* into the intended outcome.

Just as biological systems limit the effects of environmental perturbation, they also constrain the effects of mutations.* GRNs are so redundant in their construction that they can tolerate many mutations without any effect on the outcome. If process A is broken, process B

---

* These two forms of robustness may be almost the same thing; biological structures that resist the effect of environmental noise reduce the effects of mutations and vice versa. Ciliberti S, Martin OC, Wagner A. Innovation and robustness in complex regulatory gene networks. *Proc Natl Acad Sci U S A*. 2007;104(34):13591-13596.

steps in and achieves almost the same result, after which process C activates and fixes any discrepancies. Populations within a species may contain considerable "silent" variations in their non-coding DNA—and also, perhaps, their coding DNA—that do not give rise to detectable changes in developmental output and adult function.

The features discussed above apply as much to the subsystems within an embryo as to the embryo itself; the module that builds a limb builds them reliably, and it just as reliably builds toes or fingers at the ends of the limbs. Organisms are built of attractors with big basins all the way down.

Compartmentalization and the Universal Animal Toolkit

The robust but flexible system for constructing animal bodies from embryos is thought to have begun to emerge somewhere between one billion and 600 million years ago. After a period of refinement, it became a flexible, systematic, general-purpose method of constructing and varying body types. The emergence of this "universal animal toolkit" may have ignited the Cambrian Explosion, an evolutionary moment around 530 million years ago in which almost all 35 of the animal phyla that exist today rather suddenly appear in the fossil record. (See Erwin 2013 for a detailed description of this marvelous event in life's history.)

The bodies of the animal phyla that emerged during this extraordinary evolutionary moment are all built along similar lines using related and, in many cases, almost identical signaling proteins. Their body plans include a gut that goes from one end of the body to the other, taking in nutrients at one end and discarding waste at the other. The frontal area includes a high concentration of sense organs and neurons. The body is bilaterally symmetrical and constructed of three tissue layers, roughly equivalent to skin, internal organs and the lining of the digestive tract. From this basic framework, almost the entire animal kingdom emerged.

Arthropods are the most prolific animal phylum, consisting of insects, insect-like creatures (e.g., ticks, spiders and centipedes), crustaceans, and the once extremely successful but now extinct class of animals called trilobites. Current estimates suggest that there are anywhere from one to ten million species of arthropods on Earth. The phylum Chordata, comprising mostly vertebrates, comes in a distant second at 30,000 species.

The initial structures of embryos differ considerably between organisms, even within a single phylum. For example, the shelled eggs of chickens and the placenta-equipped eggs of placental mammals are quite unalike, even though chickens and placental mammals are both vertebrates. But at an intermediate embryonic stage, all organisms in a given phylum pass through a largely identical "phylotypic stage."

In arthropods (whose developmental processes are better understood than those of vertebrates), the phylotypic stage is called a germband, and it consists of an embryo of a few thousand cells divided into 50 to 60 compartments.[30] Identical protein-coding genes are present in each compartment, and many of the same genes are expressed within all of them, but their expression patterns differ. The gene regulatory network modules of the cells within each compartment operate in a compartment- and sub-compartment-specific manner. Dedicated modules activated in certain regions construct fundamental bodily features such as hearts, digestive tracts, appendages and nervous systems.

In all animals, highly conserved high-level transcription factors called Hox proteins play a major role in compartmental development, but the same signals produce different actions in different compartments and different phyla. Hox proteins go on to activate other signaling proteins specific to particular compartments or sub-compartments, such as the Pax-6 protein that initiates eye development.

This compartmental design strategy is a form of modularity: Mutations that occur in a GRN affect development in all the compartments that use that GRN, but they leave other compartments alone. This arrangement causes some aspects of an organism to remain linked and permits others to evolve independently, thus overcoming the problem of pleiotropy

For example, a vertebrate's left and right limbs share GRN patterning modules and, therefore, develop symmetrically. In contrast, the GRNs that construct eyes are (usually) activated only in the region where eyes belong; a mutation that flattens the lens will not flatten feet. When organisms come under selection, compartmentalization allows mutations to alter separate traits independently while keeping linked traits the same. In this way, compartmentalization facilitates variation, and Kirschner and Gerhart dwell on it extensively in their exposition of the theory. (For much more fascinating detail on this topic, I again recommend *The Plausibility of Life*.)

We know little about the process that created the animal-constructing toolkit, but we can only look with awe at what came next. If you will pardon another of the lyrical flights that I struggle to contain, the same basic system that built tiny worms more than half a billion years ago went on to construct animals that swim the seas, trample the land and wing through the air. The universal animal toolkit can build creatures that survive in space (tardigrades), thrive at the bottom of the ocean (numerous animals) and range in mass over 22 orders of magnitude—from rotifers to blue whales.

The toolkit achieves these things by using discontinuous switches and smoothly varying dials.

Switches and Dials

During extended periods when a trait remains adaptive, "stabilizing" selection occurs and causes that trait to accumulate redundant reinforcement mechanisms. Nonetheless, silent mutations proliferate in the control boxes of organisms in a population, including some that progressively eliminate one or more of these redundancies. These hidden GRN changes find all the available—and mutationally achievable—variant regulatory systems that lead to the same phenotypic output; they "explore" and fill out that trait's basin of attraction. Through this process, some organisms come to lie near a ridge for that trait.

If a raindrop falls in the middle of a watershed, its ultimate future is predetermined, but a mere breath of wind can shift the destiny of a drop that falls on the line of the Continental Divide; it could equally likely end up in the Mississippi or the Colorado River. Those organisms whose GRNs have come to lie near a ridge for any given trait are poised for change. All it will take is a single mutation for them to cross over the ridge and enter the basin of a different attractor.

The change produced by such a mutation may be subtle or dramatic. If a substantial module is entirely unplugged or newly connected, an organism's phenotype may undergo a spectacular and potentially useful shift in form and function. In effect, mutations like these can flip switches.

Some of the most dramatic examples of developmental switches reveal themselves in the birth defects called atavisms, such as when a human infant is born with a vestigial tail. A long-lost trait like this could only reappear in a single generation if the module used to build it remains present in human DNA, albeit switched off. Let's work through the human tail atavism in some detail to see how this might occur and what it implies.

At some point in the evolution of humans and all other great apes, tails must have ceased to be fitness-enhancing and perhaps just got in the way. Traits that cause even slight net harm become subject to "directional" selective forces that can eventually eliminate them. However, most members of the ancestral tail-possessing population would have possessed tail-forming GRNs with numerous redundant pathways. In the language of attractors, those developmental processes sat somewhere near the middle of the growing-a-tail basin. No single mutation could remove the tail trait from them.

However, the GRNs of some ape ancestors with tails would have lost all but one connection between the developing coccyx and the module that initiates tail formation; their tail-making GRNs were camped out beside a ridge on the other side of which tails are no more. For them, a single mutation affecting the last remaining control box activating the trait would suffice to shut off the tail-creating module. And, eventually, it happened.

About 25 million years ago, the tissue-specific transcription factor that initiates tail formation became damaged in such a way that it could no longer bind to and switch on the relevant control box.[31] Proto-apes with that damaged transcription factor had no tails, and the change presumably increased their fitness because it won out. However, the loss of the activating signal would not by itself disassemble the GRN(s) it once activated. Until those tail-construction modules' GRNs also decay, the ability to build a tail will continue to exist in great ape and human DNA, ready to be switched on. Several possible mutations could accomplish this: one that restores the transcription factor to its original form, another that rewires the connecting control box to recognize the altered transcription factor, or a mutation in the control box that causes it to respond to any other signal present in the vicinity. Any such mutation will produce an atavistic tail.

As it happens, humans born with what is said to be a tail possess a partial structure that merely gestures toward a tail. There are many possible explanations for this incompleteness. Perhaps tails had begun to deteriorate long before they disappeared, and the transcription factor breakdown was merely the last straw. Or it could be that multiple distinct GRNs must be activated to build a functional tail, and the atavism is produced by a mutation in a control box that activates only one. Another possibility is that tail-forming GRNs were whole at the time of the disconnection but subsequently deteriorated through genetic drift. Regardless of how the GRNs that used to build ape tails came to decay, if tails ever became fitness-enhancing for humans, directional selection could operate on the damaged GRNs that build atavistic tails and shepherd their restoration. But since humans haven't yet developed much use for tails, this hasn't happened. (We will return to the subject of the preservation of disconnected GRNs and repair of damaged ones in the section on developmental/evolutionary memory.)

This may be as good a place as any to note that using watersheds to represent the concept of basins of attraction is misleading in one important respect: Watersheds exist in a world of three dimensions, and it is only rarely the case that any given watershed has more than a few neighboring watersheds that can be reached by crossing a single ridge. However, GRNs occupy the higher-dimensional space of phenotypic possibilities. There are many ridges for each trait, each one representing a different way that trait can change; moreover, organisms have thousands or hundreds of thousands of traits. Thus, the basin filled out by all variant organisms in a population may border numerous ridges. For any trait, some members of a population are likely to be running on GRNs that have lost all redundant reinforcements for that trait and lie near a ridge beyond which that trait is altered or switched off.

So long as a trait remains adaptive, organisms whose GRNs are at the extremes of a basin for that trait will possess lower fitness than those in the middle because a greater percentage of their offspring will lack that trait; in consequence, their prevalence in the population will remain low. However, when environmental changes occur under which the trait ceases to be adaptive, the prevalence of organisms whose GRNs for that trait abut adjoining basins will increase. Since phenotypic changes are much more likely to occur via single-mutation ridge-crossing events than through changes that require two or more simultaneous mutations, the layout of phenotypic neighborhoods significantly biases and constrains what forms of variation are likely to occur. In principle, this bias could stifle future evolution; however, as discussed toward the end of this essay, the emergent geometry of phenotypic neighborhoods instead tends to facilitate it.

When environmental conditions arise such that the phenotypic effects of a particular ridge-crossing event increase fitness, the percentage of organisms in a population whose GRNs have crossed over that ridge will increase, and, in time, the entire population will consist of organisms whose ancestors crossed the same ridge. A switch has flipped. (For a more technical discussion, see Uller et al. 2018, especially pages 955-956.)

The large macroscopic changes sometimes caused by these switches provide vivid evidence of developmental modularity. If it were the case that DNA mapped organisms "pixel by pixel" (i.e., cell by cell), mutations would produce only minute, incremental changes. However, because organisms are instead coded trait by trait, mutations to control boxes can adjust and remix traits. Trait-based encoding permits organisms of trillions of cells to be produced through a system with a limited number of parameters. To borrow language from image processing, organisms are coded as vector graphics rather than bitmaps; or, for an even better analogy, as input prompts to a program like Dall-E or Midjourney.

Some evolutionary switches cause a module to be activated at a location within an embryo where previously it had remained quiescent; alterations to phenotype of this kind are said to illustrate *heterotopy*. For example, a highly conserved, widely used regulatory module builds legs when activated in a beetle's abdomen, but when it is activated on a beetle's head, horns appear.[32] Identical transcription factors trigger slightly different GRN responses in these two locations because the GRNs operating in the cells of different compartments are locked down differently. Typically, transcription factors produce the same or similar effects in repeated segments except at the terminal ends—the head and the tail.

But discontinuous developmental switches aren't the only kind of trait change seen in nature. There is a second type, too, that involves smooth, continuous change; these are the volume controls and other rotating dials or sliders of evolution. One commonly seen system of this kind flexibly alters the duration of a developmental process, causing it to occur for a longer or shorter time.[33] Toy poodles reach their full size at six months, but Great Pyrenees keep growing for 18–24 months. Mutations that alter the duration of a process are said to illustrate "heterochrony."

Other continuously variable systems utilize multiple adjustable parameters. For example, the final shape of a finch's beak is controlled by two or three transcription factors, each of which flexibly modifies one dimension of beak growth, not precisely length, depth and width, but parameters of similar effect.[34]

There are many ways that GRNs can be constructed to produce continuously adjustable traits. For example, when a control box is activated, it could initiate a process that, as one of its effects, produces rising levels of a protein product that binds to the same control box; once the level of the product reaches a set target, it switches the control box back off. A regulatory mutation that alters the sensitivity of the control box to that protein will alter the duration of the process.

In many cases, the timing of a developmental process is handled by a sophisticated, higher-level process such as the release of endocrine hormones under the control of a developing or mature central nervous system. Because these hormones function merely as signals, the same or similar hormones may be used for many different purposes in different creatures. Thyroid hormone, for example, causes larval amphibians to mature into adults, plays a role in human puberty and regulates elements of embryonic brain development in many animals.

When heterochrony goes far enough, it can dramatically change organism form. For example, bats evolved from animals with more typical hands by extending the growing period of their middle fingers so that they grow longer.[35] Fingers are natively supplied with webbing, but in human embryos, a module comes online that causes the webbing between fingers to self-destruct. Bats retain the webbing and extend it.

The metamorphosis process that produces butterflies from caterpillars is a more sophisticated (and jaw-dropping) example of heterochrony. Non-metamorphosing insects such as grasshoppers emerge from their eggs as small, perfectly formed adults. However, during an intermediate stage of their development, grasshopper embryos look something like digestive tracts with legs, a description that also applies rather well to caterpillars. A subsequent surge of hormone-induced modifications causes some elements of this pseudo-caterpillar to be absorbed and others added, yielding the adult grasshopper.

This process is altered in insects that first emerge as larvae (caterpillars) and reach adult form (butterflies) only after building a cocoon. When the developing butterfly embryo reaches the digestive-tract-on-legs stage, it hatches "prematurely" and wanders around the external world, devouring leaves. After a month or so of living life as a walking embryo, the delayed hormone surge occurs and triggers cocoon formation. The within-cocoon conversion of a caterpillar into a butterfly is simply a delayed version of the developmental process that originally occurred inside the egg.[36] No new stage had to be invented to innovate the egg/larvae/cocoon/adult lifestyle from the egg/adult one. Many larval stages observed in animal life cycles seem to have originated as forms of delayed embryonic development.[37]

The developmental switches and dials available in the animal toolkit are sufficient to produce the great range of variation that I have periodically rhapsodized about above.

Heterotopy, heterochrony and many other processes can produce dramatic changes in organism form, such as expanding a Chihuahua to Great Pyrenees size or giving it the flat face of a Pug. However, one can easily imagine that a significant alteration to elements of a body plan would create problems for subsidiary systems and adjacent structures. For example, if heterotopy builds a third eye in a reptile (true story), other processes must come online to supply blood to it. One might imagine that this would require a (highly unlikely) set of near-simultaneous mutations. But that is not the case. Instead, evolution has acquired a capacity to reduce its dependence on supportive genetic changes by using "smart widgets," flexible developmental tools that fill in missing pieces on their own rather than waiting for slow evolutionary processes to do all the work.

Exploratory and Adaptive Processes

GRNs resemble computers, but they are not particularly smart; amoebas possess flexible behavioral and physiological processes controlled by GRNs, but their within-life adaptive powers pale before those of organisms with even rudimentary nervous systems. Over evolutionary time, GRNs were shaped by evolutionary processes in such a way that they came to build structures and processes capable of autonomous, "intelligent" behavior. Brains are the most sophisticated of all such adaptations, but organisms also utilize other systems whose independent capacities supplement the computational power of GRNs. Among these are various *exploratory* and *adaptive* processes that solve developmental problems on the fly. These inbuilt capacities constitute another primary element of Kirschner and Gerhart's theory of facilitated variation. Their ability to improvise greatly reduces the number of mutations that would otherwise be necessary to successfully modify an organism. Evolution has learned to delegate.

For a classic example of an exploratory process, consider the mechanism that builds circulatory systems during embryonic development.

To support complex multicellular organisms with more than a few cell layers, evolution had to find a way to supply internal cells with oxygen. The system that emerged uses a pumping device that circulates an oxygen-rich liquid through vessels arranged in such a way that no cell is separated from the nearest vessel by a distance that exceeds about a tenth of a millimeter.

Major arteries exiting the heart subdivide into smaller ones that go on to further subdivide. Although vessels can actively expand and contract to a certain extent, the average diameter of each vessel must generally correspond to the average downstream oxygen needs. It would take a considerable act of information gathering and computation for a centralized system to design an optimized arterial system for each new organism form, but embryonic development manages it through a simple algorithm.*

A few large arterial vessels appear early in embryonic development and are probably pre-programmed (somehow) in the GRNs that

---

* The same system continues to operate throughout life and helpfully provides cancerous growths with their oxygen needs, too.

control the organism's development. But spontaneous processes fill out and calibrate the rest of the circulatory system. As the embryo develops, sprouts emerge randomly from the existing arterial vessels' walls and explore their surroundings. Tissues that are inadequately supplied with oxygen because there are no vessels in their immediate neighborhood produce signals that draw angiogenic (blood vessel-creating) cells into the gap between oxygen-starved cells and the nearest sprouts. These cells spontaneously link up to form new blood vessels that bring oxygen to precisely where it is needed. In addition, friction from moving blood stimulates vessels to increase their diameter. Those vessels with the greatest downstream demand experience the greatest friction, causing them to enlarge the most, while vessels with low flow rates remain small or shrink. This distributed process of exploration and reinforcement efficiently matches blood vessel size to downstream oxygen needs without any centralized control.

Veins emerge in much the same way, although they are stimulated to grow by the waste products of cellular respiration rather than oxygen demand. You can find evidence of this improvised design by rolling up your sleeves and examining the asymmetrical superficial veins of your right and left forearms.

The nervous system, too, develops through exploration. Neurons in the spinal cord devoted to muscular control send out long extensions (axons) that subdivide into fine branches and proceed to seek out muscle cells. Multiple axons contact each muscle cell, but a subsequent process eliminates all but the single axon best placed to most efficiently stimulate it. This competitive system yields an efficient peripheral nervous system design.

Yet another exploratory process guides muscle development. Signals given off by developing bones attract wandering progenitor cells that settle nearby and proceed to form muscles and tendons. If researchers transplant the cartilaginous precursor to the bony pattern of a limb (a limb bud) anywhere in an embryo, muscle cells will migrate there and try to build a muscle system to operate it.

On-the-fly processes are not limited to animals or even to multicellular organisms. Kirschner and Gerhart use the self-assembling internal cellular cytoskeletons of amoebas as their premier example of the power of exploratory processes (Kirschner and Gerhart 2006, Chapter Five).

Exploratory processes use local rules to produce global benefits. In this they resemble the algorithms used by ant colonies to efficiently exploit food sources. Although each individual ant follows a mere handful of rules, when hundreds of thousands of ants follow the same rules, the result resembles the action of a centralized thinking machine.

In addition to exploratory processes, organisms also employ adaptive processes that cause anatomic structures to automatically adjust to the physical forces that act on them. One of the most dramatic of these is the mechanism that can cause a joint to emerge spontaneously in a region of developing bone when certain forms of physical tension are consistently applied there.[38]

Bones themselves are largely preprogrammed in their initial formation, but their structure responds faithfully to the strains they experience. For example, persistent linear compression causes bone density to increase and bone structure to refashion itself along the lines of arches and corrugation. Conversely, when average strain decreases, bones rapidly thin; this is one reason why astronauts living on the International Space Station must exercise daily. Muscle growth in response to exercise is another such process and a second good reason to exercise when spending extended time in zero gravity.

Exploratory and adaptive processes pave the way for phenotypic evolution. If nerves, muscle, bone and blood vessels each had to be adjusted separately, significant physical changes would require many correlated mutations in a lengthy do-si-do fashion. However, because exploratory and adaptive processes work out many details on their own, much fewer mutations are needed. Exploratory and adaptive processes are exceedingly smart widgets; like skeleton keys, they unlock many adaptive doors.

The power of these systems reveals itself quite remarkably, although usually without remark, in dog breeding.[39] Suppose I appreciate a Pug's personality but prefer the long snout of a Whippet. I don't need advanced genetic engineering to accomplish this; I can simply breed one with the other, and not a single member of the resulting litter will emerge from the womb with an upper jaw elongated like that of a Whippet and a lower jaw snubbed like a Pug. Instead, the offspring of such a match reliably resembles a cross between both parental breeds, mixing or averaging their characteristics. We take this outcome for granted and fail to recognize its profound significance.

If a GRN for a longer upper jaw from the Whippet parent has been introduced into the Pug/Whippet combination, one might suppose it would need to be accompanied by GRNs that extend the lower jaw, tongue and accompanying muscles, nerves and blood vessels. Given that chance plays a role when parental traits mix during sexual reproduction, one wouldn't expect every puppy in a litter to possess all the necessary matching elements. Nonetheless, crossbreeding reliably produces coherent dogs rather than monstrous and perhaps lethal collages of unmatched pieces. This successful admixture results, at least in part, from exploratory processes that fill in details and smooth out contradictions.

But one doesn't need to consider deliberate breeding to see such things; similar processes cause children to generally resemble both their parents rather than emerging from the womb as bizarre Cubist montages of parental traits.[40]

If human engineers could build things using exploratory processes, the manufacture, fabrication and construction of hybrid products would become much easier. For example, suppose the buyers of Jaguar coupes inexplicably also wanted their cars to have the boxy look of a Kia Soul or Jeep Wrangler. With enough time and effort, automobile engineers could design and build a car that combined those features. However, they couldn't simply merge the assembly lines of the two or three types of vehicles. They would have to build an entirely new assembly line (or significantly modify the existing ones, and design and manufacture subsidiary parts, such as exhaust pipes that suit the new vehicle.

In contrast, a manufacturing system employing exploratory processes to grow components and systems on the fly would handle design changes much more effectively and efficiently than a fixed assembly line that builds to a blueprint. (It would also be creepy. We trust that the growth processes in living organisms are well-behaved but might fear the possibility that rogue exhaust pipes might grow off the assembly line, head to the canteen and wrap around the legs of employees who linger too long at lunch.)

Systems that follow general principles but permit local, spontaneous improvisation can be especially flexible, robust and well-suited to solving novel problems. For this reason, the military doctrine utilized by most modern states grants units on multiple sublevels the freedom to improvise their responses to local conditions. Experience has shown that when leaders of squads, platoons, companies, battalions and brigades can be trusted to choose their own solutions to the problems they face, the army of which they are a part becomes more effective; conversely, rigid top-down control makes a military brittle and inefficient.[41]

Improvisation within general principles can be viewed as a form of generalization. Flexible military doctrine is a generalization over varying battlefield conditions, and the flexible exploratory and adaptive processes described in this section are generalizations over the task of building and operating various animal bodies. Exploratory and adaptive processes facilitate evolutionary change by smoothing over the results of DNA mutations that might otherwise be lethal, minimizing the production of monsters by adaptively combining traits and decreasing the number of mutations needed to produce a substantial, adaptive change.

Remarkable as they are, exploratory and adaptive processes are not the only general-purpose, evolvability-enhancing tools that evolution has discovered.

Skeleton-Key Widgets

It is a common criticism of surgeons that they are too surgery-prone—a tendency captured in the expression, "To a person holding a hammer, everything looks like a nail." However, many things look like nails in one respect or another and hammers are useful for many purposes; they are a generalization over common tasks. And so are nails. Artisans invent reliable and broadly useful tools through long experience of making things, and so does evolution: The tools, structures and processes used in the construction of animal bodies can be counted on to work well in a range of settings.

The exploratory processes described in the previous section are robustly effective tools for embryonic development that work well for any body plan, and the completed body of an organism contains numerous broad-spectrum tools or widgets directed at the external world. Some of these are so widely useful that they can be thought of as skeleton-key solutions to the problem of survival, such as eyes, ears, fur, feathers, jaws with teeth, limbs with claws and nervous systems that learn by association. As described in the theory of facilitated variation, evolution typically proceeds by recombining and tweaking existing traits rather than inventing entirely new ones. What gives this system for creating organisms much of its power is that many of these traits are profoundly general purpose in their capacities.

Evolution has learned how to build tools that tap into persistent regularities of Earth conditions. Skeleton-key widgets are generalizations about the world. If converted to explicit propositions, they would read something like this:

- "Sharp objects can cut, tear and rip things." This is a generalization over many materials.
- "You can build an excellent arterial system by responding locally to oxygen deprivation and total blood flow." This is a generalization across numerous body plans.

- "The effectiveness of a sharp object is enhanced when it is attached to a powered lever, and fins or limbs do a better job when there is one on each side of an organism's body." These are generalizations about forces.
- "Objects and events of relevance to survival can often be detected by sensors tuned to certain frequencies of light." This is a generalization across objects and events.
- "Neural networks built up into brains can retain information, learn associations, find patterns and solve novel problems." Neural network architectures of sufficient complexity can produce useful generalizations about so many phenomena, and they currently remain so far from having reached their limits that the range and limitations (if any) of their generalizing capacities remain unknown.

Some biological processes are tuned to a single goal. For example, the enzyme systems that manage fundamental metabolism are fantastic at doing that but not much good at anything else. Similarly, the red crossbill's peculiar beak is optimized for feeding on pinecones and the like, but it impedes them from consuming other foods. However, most biological processes are tools of many uses, and, moreover, they can be relied on not to jam, misfire or break on their second use.

Many of the processes and traits that first emerged in evolution for one set of purposes turned out to be good for many others. For example, language did not originally evolve as a system for formalizing logical propositions, but it works well for that and many other novel purposes. At a less exalted level, fins evolved for propulsion through water, but with slight modifications, they also proved effective for walking on land.

Such extensibility of use should not be surprising. Because general-purpose tools exploit universal or at least common regularities found in the world, they are likely to prove useful for tasks other than those for which they were first designed; such novel tasks can be said to possess structural similarities to prior ones. Hammers are not only good for nails but for rivets, too.

For a biological example, consider actin microfilaments—microscopic threads composed of multiple copies of the protein actin linked together. Microfilaments probably first emerged to serve as a kind of cellular tether, holding structures in place and sometimes exerting small amounts of force on them. However, evolution subsequently found many other uses for actin microfilaments. Some of their most important new applications arose when actin microfilaments teamed up with another protein, myosin, to form contractile fibers; eventually, these would become the basis of muscles.

At a fundamental level, actin microfilaments are broadly useful because they reflect Newton's second law: that the motion of matter depends on the sum of the forces acting upon it. Filamentous structures exert forces linearly, and although they are not of much use for pushing, they can pull objects or tether them in place. Such microfilaments are broadly useful because they correspond to a universal regularity of the physical world. Their evolved physical structure is itself a form of knowledge; when evolution "invented" and refined microfilaments, it acquired a kind of analog understanding of Newton's second law.

Many of evolution's widgets reflect *identifiable* features of the physical world, and these correspondences make it easy to understand why they are useful for multiple purposes. Feathers may have first evolved to retain heat, but the same characteristic that makes them so useful as insulation—their ability to impede the movement of air—also adapts them for flying. Eyes respond to frequencies of light that provide considerable information about many events of relevance to mobile organisms. Symmetrical legs or fins produce balanced forces that contribute to linear motion in most environments. Legs with multiple joints permit stabilized perambulation on many uneven surfaces. Contractile cells operating on partially enclosed containers can propel liquids and gases. And, because neural networks can produce models to fit (almost) any data source, they can predict many events and solve a great variety of problems.

However, there is no reason to think that the features of the world that are cosmically universal (like Newton's laws) or that are less cosmic but that our minds easily recognize (like the value of articulated legs) encompass everything of relevance to the survival of organisms. Most biological tools likely operate as generalizations over events, circumstances and conditions that we know nothing about. No one would have recognized that a particular bent wire shape would prove nearly optimal for holding sheaves of paper together until paperclips were invented (and until bundling such sheaves became a necessity of life); similarly, there is no doubt that many biological features address recurrent challenges of which human observers remain in ignorance, such as the particular materials and techniques that make it easier to catch insects by extending a long, sticky tongue. But because it is easier to talk about traits that our intuitions encompass, I will usually talk about legs and fins rather than the adhesive characteristics of sticky mucous and the aim-ability of variously designed tongues.

Other than actin and myosin, I am also neglecting those innumerable widgets of great power and generality that operate on a microscopic level, such as the ancient systems that can preferentially choose which ions pass through a cell membrane or inactivate toxic molecules. Still, similar principles should drive the emergence and refinement of skeleton-key widgets no matter at what scale they appear, and, hopefully, examples chosen mostly from visible anatomic traits can successfully stand in for the rest.

Once a general-purpose widget of any kind has been invented, evolution can modify it for different but related uses. Materials scientists can explain that sharp objects are generically useful because they concentrate compression forces in such a way as to break molecular bonds. Evolution has discovered this principle and elaborated teeth, claws, beaks and spines of various sizes and shapes. Size modification is itself a general-purpose tool that acquires its generality from the mathematical properties of linear relationships: If object A is five times bigger than object B, the laws of its behavior are the same as those of object B, only at a larger scale. Here, "size" can mean such parameters as length, surface area or volume. Scaling may also be useful in some non-linear circumstances because of the fractal character of the world: Relationships between structures and their substructures often remain the same on larger and smaller scales.

A tool will sometimes acquire entirely new properties when its design is stretched beyond a certain point. For a somewhat disturbing example, consider the origin of external cheek pouches in rodents. Many rodent species possess *internal* cheek pouches that allow them to stuff their faces with several weeks' worth of food. However, pocket gophers and kangaroo rats possess fur-lined *external* cheek pouches. The origin of these pouches had been something of a mystery until researchers discovered that when the developmental processes that construct an inner pouch are modified beyond a certain point, a topological transformation occurs that causes the pouch to develop on the outside of the body.[42]

In addition to modifying, tweaking and optimizing widgets, evolution can combine them. Limbs and jaws are useful in part because they operate as levers. Add a claw or tooth to a powered lever, and you have something that resembles an axe or a wedge. Add a muscle-controlled tail to a four-legged body and you have a creature with superior balance. Combine one forward-facing eye with another and you have efficient binocular vision.*

The modular, feature-based design of organisms makes it easy for evolution to combine traits, rather like an image-generation program can successfully respond to prompts such as "Draw me a giraffe with butterfly wings, a unicorn horn and a Groucho Marx disguise." Programs such as Dall-E or Midjourney can easily draw compound objects because they have decomposed images into features. Evolution hasn't

---

* Surprisingly, animals whose eyes are located on the sides of their heads, such as horses and rabbits, can also extract distance information from the environment; they use their visual neural networks to compare the images they see when they move their heads from side to side.

created those organisms precisely, but it has produced plenty of equally weird designs, such as giant pigs with tentacles (elephants), reptiles with armor-plating made from fused ribs (turtles), and limbless ocean-going mammals that eat plankton strained through filters (whales). One can be sure that if a Groucho Marx disguise were recurrently useful, it would evolve, too. It would make a stylish butterfly wing design.

Add the "grow hair" module to "build a tail," and you get a horse's tail. Leave out the hair, and you get the tail of an opossum. Turn on "build a bone" in the skin and, voilà, you have the plate armor of an (extinct) armored fish. Turn "build a callus" up to 11, and you end up with a scale. Combine two rods with a joint and you get an arm that bends at the elbow. Add another joint, and you have a wrist. Keep going, and you have fingers.

Just as amino acids are built out of atoms, proteins out of amino acids, and fibers out of repeated proteins, animal forms are constructed by combining simple features into compound features and then combining those compound features into higher-level ones.

Evolution has thus acquired general-purpose widgets that solve a range of problems as is, can be tweaked to solve even more problems and can be profitably combined with other widgets to solve a still wider range of problems; the animal toolkit thus includes a tool for just about anything life can throw at them. The fact that general-purpose widgets facilitate future evolution without coming under selection for that purpose is not mysterious; because they achieve their general-purposeness by tapping into fundamental characteristics of Earth environments, they will tend to be useful in numerous previously unsampled environments too.[†]

However, the same question arises here that came up in the discussion of pluggable modularity. Recall that non-modular hacks are frequently more effective in the present and should, therefore, often emerge under

---

[†] More subtly, tools that tap into deep structural regularities of the world permit evolution to proceed further along directions that exploit those regularities.

selective processes; the near-universal presence of biological modularity therefore requires explanation. The analogous question here asks how it is that evolution has discovered and retained skeleton-key widgets that tap into deep characteristics of the world rather than tuning up a host of specialized tools best suited for current conditions?

This question is much easier to answer.

How Has Evolution Found Its Skeleton-Key Widgets?

Evolutionary processes (and human engineers) are frequently compelled to find compromises. Flat claws are best for digging and pointed ones for stabbing, but it would be hard for animals that engage in both behaviors to bear two sets of claws. Optimal compromises emerge spontaneously under direct natural selection and naturally yield multipurpose solutions such as claws that are reasonably useful for multiple purposes. The necessity of compromise is plausibly the primary force that drives the evolutionary invention of general-purpose adaptations.

If selection pressures at the moment bias adaptations toward general usefulness, repeated bouts of selection continue the process. Some general-purpose widgets may resemble slide rules and old-fashioned cordless phones in that they are useful for relatively brief historical periods and then disappear. However, others, like wheels and nervous systems, never go out of style. Adaptations that solve a great many problems and continue to be useful in a wide variety of conditions are the crème de la crème of widgets, and these are the ones best placed to survive sequential selective environments, much as the smallest grains of sand are all that's left when raw sand passes through a series of sieves. Such iterated direct selection will plausibly bias evolution toward discovering, retaining and progressively optimizing skeleton-key widgets. For a human example, consider metal nails. Ancient Egyptians used bronze nails as long ago as 3400 BC, and nails remain in use today because, as age followed age, they remained the best tool for many tasks.

Widgets may also persist if they are constructed and produced in such a way that they can be readily tweaked to match changing conditions. Examples include beaks and teeth, which, as discussed later in this essay, are produced through parameterized developmental processes that permit rapid evolutionary modification.

Iterated direct selection combined with continuous tweaking might explain the origin of exploratory and adaptive processes. It is not easy to see how these could have emerged as compromises; what would they have been compromises between? However, they could plausibly have begun as limited-scope, local processes that continued to be expanded and optimized because they remained useful even as conditions and organisms changed. Writing is an example from the human world. It may have originally been used only for specialized purposes, most prominently trade and commerce, and for many centuries it must have been solely the province of specialized scribes, but because it was such a useful and flexible tool, its uses and the range of people who used it continued to expand.

Iterated selection may also tend to eliminate specialized, limited-use widgets because the conditions for which they are optimal do not persist. One can easily imagine that this will be the ultimate fate of the dangling fake bait used by anglerfishes. Pruning of this kind may help prevent the dead hand of the past from too severely constraining future evolution: those widgets that survive iterated selection offer benefits (or can easily be tweaked to offer benefits) under a great variety of Earth conditions This process of pruning can be understood as a form of implicit induction: Adaptations that are broadly useful in the present and continue to be broadly useful under changing Earth conditions are likely to remain useful in future conditions, too.

Interestingly, lingering influences from the past might sometimes *facilitate* the emergence of skeleton-key widgets by permitting them to endure through rare, brief periods when they are not adaptive. This persistence is a kind of action against selection, and it plays a role in Watson's theory of natural induction through self-modeling, too.

I will readily admit that the arguments in this section are a bit hand-wavy. To make them more rigorous, one could run computer simulations of virtual organisms possessing virtual traits and track what happens as their virtual environments change. Not coincidentally, that is very much how Watson and his colleagues have gone about verifying their ideas.

A Brief Introduction to Induction, Generalization, Extrapolation and Modeling

In these pages, I have been tossing about such terms such as induction, generalization and modeling, and will do so even more in what follows. Let's pause the narrative and give them working definitions.

To acquire useful knowledge of the world, we sample events and look for patterns. Once we find a pattern, we examine a larger set of events to see if the pattern still holds. If it does, we come to believe, at least provisionally, that the pattern is universal. This is the process of induction, or inductive generalization: We predict the future by generalizing from the past.

A well-characterized generalization is a kind of informal model. When a recurring pattern in nature attracts the interest of scientists or engineers, they will attempt to build a formal model that represents the generalization mathematically. Here are some examples of these stages of induction:

- Observation: That apple fell, and so did two others on the same tree.
- Simple inductive generalization: All the apples on this tree will eventually fall.
- Further inductive generalization: All objects held above the ground and dropped will fall.
- Informal model: Objects accelerate as they fall freely. The farther they fall, the faster they go; it doesn't take long for them to be moving dangerously fast.

- Formal model (with additional insights added): Neglecting friction, the motion of a freely falling or flying object can be decomposed into two vectors, one pointing toward the center of the Earth and the other perpendicular to it; the magnitude of the radial vector in meters per second is 9.81 meters per second per second times the elapsed time in seconds.

Models are generalizations built by studying subsets of a larger data set and applying them to the whole. Newton deduced his law of universal gravitation by studying the motion of the Moon and planets, but we confidently expect that the same laws would apply to the moons and planets of other solar systems, too. On a more practical level, NASA scientists relied upon the same laws to predict the motion of spacecraft. Good models built on data subsets often have something meaningful to say about unsampled data, too. That is what we mean by generalization.

Models that have proven effective in many and various situations beyond those that led to their original creation may elevate their inventors into scientific heroes. The generalizations embodied in Newton's laws have turned out to be so widely valid that he was buried in Westminster Abbey and immortalized by Alexander Pope with the somewhat less-immortal lines "Nature and Nature's laws lay hid in night: God said, Let Newton be! and all was light."

Around 1.7 million years ago, our protohuman ancestors learned how to make "hand axes," stone tools with a cutting edge on one end and a thicker surface on the other shaped for easy gripping. They found this tool so useful that they used it without change for more than a million years. A protohuman setting out with a hand axe is engaging in a form of inductive generalization over tasks. If converted to a proposition, it would state, "What helped me yesterday is likely to help me today." Evolution, too, generalizes over tasks by using such general-purpose widgets as exploratory processes, muscles, hands, eyes, claws and brains.

Sharp tools like axes and claws are generally useful because they implicitly model a persistent characteristic of the world: When force is concentrated, it can cut things. As organisms evolved progressively more sophisticated nervous systems, they acquired the ability to create progressively more complex models. Learning to ride a bike consists largely of giving the neural networks in the cerebellum (a substantial sub-unit of the brain that calculates movements) sufficient time and the inductive knowledge acquired through successes and failures to build an effective model of bike riding. The completed model takes input from the eyes and inner ears and outputs well-calculated responses in the form of steering, leaning and pedaling to produce successful forward motion with a minimum of crashes and falls. When visual data arrives showing the world taking on a rightward tilt and data from the inner ear confirms the impending loss of balance, the cerebellum will calculate the appropriate response to avoid a fall and transmit the results of its calculations to the upper motor neurons of the cerebral cortex. Once the cerebellum has constructed a model, it holds on to it; hence, the common experience that no one forgets how to ride a bike.

Although the cerebellum's bike-riding model is built of neural interactions rather than written out in formulae, it is no less a formal model than the laws of Newtonian physics. It addresses much the same range of phenomena and differs mostly in that it uses analog techniques rather than symbolic calculus to solve differential equations.

When we make an analogy between one idea and another or say that two conditions or processes possess structural similarities, we are suggesting that the model used for one problem can profitably be applied to the other. If the analogy is good enough—the structural similarity is sufficiently great—we may be able to use our mastery of the first problem to expedite the solution of the second. People who can rollerblade proficiently possess a cerebellar model that can handle the basics of ice-skating and skiing; surfers one that works fairly well for snowboarding.

The neural networks that constitute brains are particularly well suited to modeling, but biology does not limit itself to brain-based models. The attachment of fins to a rigid spine in effect models a class of solutions to the challenges faced by creatures whose lives depend on their ability to generate rapid and precise motion through water. When fish moved onto land, they encountered novel conditions for which their previous model for propulsion was an excellent analogy, requiring only the modification of fins into limbs.

When we have created models for a variety of similar experiences, we may go on to develop a model of models, a generalization across models that produces specific ones as needed. For example, when reasonably sensitive people travel to a country they have never visited before, they gradually learn how to behave properly according to the customs and manners of that country; stated more formally, they construct a country-specific behavioral model. If they travel widely, they gradually build a model of models that allows them to create country-specific models efficiently. An experienced traveler can adapt to the customs, practices and habits of a new location much more rapidly than someone who has never left home. Similarly, AlphaGo, the deep learning program that beat the best human Go player, mastered chess rapidly because the model it used to create models was also able to model chess.

Embryonic developmental systems are models of how to build animals. Because widely varying animals are constructed with the same toolkit, that toolkit is a model of models or a generalization across models. When directionless mutations occur, they, in effect, alter parameters in the general model and cause it to model a modified organism.

The theory of facilitated variation can thus be expressed in the language of models like this: The model of models that has emerged over evolutionary time and that resides in the GRNs that map genotype to phenotype optimizes evolutionary variation by modeling the construction of novel organisms responsive to altered conditions. The next logical question might be, "What processes optimize the genotype-phenotype mapping function?" Direct selection plays a role, but, according to Watson, so does the emergent process of natural induction.

Optimization of Optimization, of Optimization

To optimize is to produce the best possible outcome within a set of constraints. It's (almost) never possible for us to get everything we want because some of the things we desire conflict with others, as is commonly apparent in household budgets.

For an example more relevant to biological design challenges, consider the issues that complicate the design of an all-season tire.

Tires with no tread grip extremely well on dry asphalt but will hydroplane in wet conditions; tread patterns that are ideal for water are less than ideal in snow and mud and always worse than no tread at all for smooth, dry roads; tread patterns that work best in snow and mud are problematic in icy conditions; softer tires grip better under all conditions but will rapidly burn up in summer. An "all-season" tire is a generalization over year-round driving; it provides an average performance on a year-round basis that would be superior to how a summer tire or a winter tire would perform over a year, but it can't match the performance of specialized tires used in the specific conditions they were designed for.

Even if we wanted to optimize a tire for a single condition, we would run into conflicts; for example, it isn't possible to simultaneously optimize a tread for initial excursions into the snow and consistent, effective traction once the snow gets packed in. Engineering involves compromise all the way down.

Evolution faces similar problems. Larger wings give greater lift, but they are heavier. Brains enhance many capacities, but they suck up calories and are vulnerable to injury. Evolutionary adaptation inevitably comes up against the need to compromise.

Brains, too, spend most of their time calculating compromises. As the cerebellum refines and improves its model of how to successfully ride a bike, the owner of that cerebellum begins to react more optimally to such challenges as grates, bumps, potholes and cars backing out of driveways. Nonetheless, every bike-riding decision must operate within the constraints of the bike's structure, the laws of physics, the time required to complete the calculations, the limits of the speed of nerve impulses and the concatenation of multiple incoming risks.

Optimization processes typically act on many different timescales. Darwinian evolution is a slow-acting optimizer, and the results of its optimization processes are the traits we call adaptations. Brains and nervous systems are fast-acting behavioral optimizers created by slow-acting Darwinian processes.

Optimizations on different timescales typically affect one another. For example, brains have been optimized by Darwinian selection so that they can most effectively guide the within-lifetime behavior of organisms, but, as discussed in a later section on the Baldwin effect, brain-guided behavior can return the favor and direct future variation in profitable directions. Thus, a product of Darwinian processes can optimize those processes. While this bidirectional optimization might seem remarkable, it's actually ordinary. To see it in action, let's examine the common process of learning at school.

Suppose serious high school graduate Georgina matriculates to college. When she takes her first quiz, she draws on all the resources of her native intelligence and her high school-educated brain when she chooses her answers. Test-taking, like organism survival, involves optimizing behavioral responses in a demanding situation where resources and information are limited. Georgina is an excellent test-taker already. She visualizes her notes when a detail seems out of reach, keeps track of the time and paces herself, chooses the least bad answer to multiple-choice questions by making the most of the facts she remembers along with more general logical analysis, and when answering essay questions, she sticks to what she knows and avoids making things up or merely filling space with words.

However, she can improve her test-taking skills by studying. Studying is a behavior that optimizes test-taking behavior. The more Georgina studies, the better her brain can operate during her next test, more frequently choosing the correct multiple-choice answer and writing essays along the lines of what her professors are looking for. But there is also an influence in the reverse direction: Each test she takes gives her information on how she should study for the next test. If she often forgets facts, she memorizes facts more intensively. If what she lacks is memory of categories and relationships, she focuses on those elements of knowledge acquisition.

As she takes tests over the course of the first semester, Georgina becomes better at studying. She discovers when to keep pushing and when to take breaks; she discovers methods such as rewriting notes and using flash cards. She is optimizing (by revising her study habits) an optimizing behavior (studying) that improves her test-taking success (itself an optimization process). But here, too, there is also an influence in the opposite direction; her successes and failures teach her how to study better. If a certain technique helps her memorize Japanese vocabulary faster than another, she will rely more on that method. There is also an influence from the lowest level of optimization to the highest because her test-taking performance is the ultimate verification of the efficacy of her evolving study methods.

Next suppose that she has taken only one language class to date, Japanese, but now she enrolls in a Chinese language class. Having learned Japanese, she knows something about how she learns languages. She knows what aspects of language learning come easy to her and what aspects she struggles with. She has also learned the optimum proportion of verbal versus written practice. She has optimized the optimizer of an optimizer of an optimizer.

Or, to switch to the partially equivalent language of models, she has produced a model of how she learns languages that itself models the effectiveness for her of various studying styles, each of which is a model of how to acquire certain elements of knowledge required to properly choose the right mental model when answering test questions. She has learned how to learn.

What we call *practice* is the optimization of an optimizer that may itself be an optimizer or an optimizer of optimizers; it is the building of a model for world-traveling that models how to model adaptations to novel countries; it is a generalization over multiple sessions of rollerblading that produces a system for navigating on non-gripping shoes that is so general-purpose it also works for ice skating.

Evolution also practices. However, there is a by-now familiar difference between the layered optimization of a human student and the various forms of optimization that are accomplished by natural selection operating on variation. Human brains have *evolved* to pursue short-, medium- and long-term goals, and it is this capacity and drive that causes us to improve our models at many levels and timescales. Natural selection, however, is trapped in the present. Nonetheless, the past often predicts the future, and past evolutionary experiences have left a persistent mark on DNA sequences and the computational processes they encode. These lingering effects act as a model of past models and can preferentially offer up adaptive variations in the face of novel challenges.

Three Primary Timescales of Biological Optimization

A student spends an hour taking a test, many hours preparing for tests, multiple semesters becoming better at preparing for tests and years becoming better at learning how to learn. Evolution, too, optimizes at many timescales. I will focus on three, but the division is not meant to be precise or fundamental. Each timescale can be meaningfully subdivided, and the extreme range of each scale may overlap the extreme range of the one below or above it. Nonetheless, these three approximate timescales adequately represent three fundamental categories of evolution's optimizing processes.

Timescale One ranges from fractions of a second to several generations. Optimization on this timescale consists of an organism's within-life capacity to respond as best it can to current circumstances, augmented by "epigenetic" processes that can extend such responses for a limited number of subsequent generations. Taken together, these phenomena are forms of phenotypic optimization or phenotypic adaptation; they constitute the adaptive capacities possessed by creatures with a given GRN/gene endowment. The phenotypic adaptations available to an organism function as generalizations over the experiences that organisms are likely to face within their lifetimes or within a few generations.

Ordinary Darwinian selection, the traditional topic of evolution, operates on *Timescale Two*. This optimizing algorithm typically takes a couple of million years to produce new species in plants and animals, although there is no fixed rule, and speciation can sometimes occur far faster than that.[43]

Darwinian processes literally incarnate induction; they select organisms with the highest current fitness and, in effect, bet the genetic farm that whatever works well today will continue to work at least to some extent tomorrow. This method can certainly fail: Ask the trilobites that once dominated the seas yet left no descendants. However, it frequently succeeds. If it didn't, there would be no living organisms because none would have survived changing conditions long enough to adapt to them.

Timescale Three is the period over which evolution itself becomes optimized. For example, the universal animal toolkit—with its compartments, pluggable modularity, exploratory processes and other gifted widgets—came together not over millions but hundreds of millions of years. This system and its subsequent refinements facilitate evolutionary change by influencing variation; as I will argue in the last third of this essay, the reason they can do so lies not so much in the blind, driving processes of natural selection but in the parallel process of natural induction through self-modeling identified by Watson.

Just as with test-taking and studying skills, the various timescales of biological optimization affect one another. One of these relationships is obvious: Darwinian selection on Timescale Two continually optimizes the within-life adaptive capacities that operate on Timescale One. That's what we mean by "evolution." But there are optimization relationships in other directions also. These will be discussed in the next several sections.

We will begin with a discussion of phenotypic adaptations on Timescale One and then show how those reach back up to optimize processes on Timescale Two. The interaction between Timescale Two and Timescale Three processes is more subtle and will occupy the final third of this essay.

Optimization Timescale One: Phenotypic Adaptation/ Plasticity

Living organisms are not static; they adapt and respond to environmental challenges. When pushed to the edges of their comfort zones by shifting environmental conditions, they respond in ways that restore homeostasis. Some physiological responses occur within fractions of a second, while others deploy over months or even years, but all occur at rates much faster than the median rate of Darwinian evolution.

These adaptive responses are intelligent choices in that they are well-calculated to enhance survival. Evolutionary processes operating on Timescale Two built (almost) all the adaptations used by simple organisms these include the hardwired responses we call instincts. However, as increasingly complex nervous systems came online, learned experience began to contribute a greater share. Brains are to GRNs as computers are to humans; they offer a kind of outsourcing and amplification of abilities already possessed.

Nervous systems are (probably) the most rapidly acting macroscopic response mechanisms that evolution has created, and all animals except for certain sea sponges use them. Examples of homeostatic actions mediated by nervous systems include the tendency of mammals to sweat, pant and flush when they are hot and to shiver, elevate their fur and move about more rapidly when they are cold. Other inbuilt physiological processes operate wholly or in part without nervous system intervention, such as tanning after UV exposure, increasing the total number of red blood cells in response to the reduced oxygen levels at an elevation of a mile above sea level, building up a resistance to a repeatedly encountered infectious agent and growing larger muscles in response to exercise.

Some adaptive responses operate by altering gross anatomy. Among animals, this is especially common in insects. For one example among thousands, water fleas (*Daphnia cucullata*) are sensitive to the chemical traces of predatory damselfly larvae, and when they detect such traces, they grow large helmet-like heads that make them more difficult to eat.[44]

Plants, too, can respond physically to threats. When certain species detect that insects or deer are extensively damaging their leaves, they respond by producing toxic or noxious chemicals, increasing the size of the spines that grow along their stems, or generating signals that attract the predators of whatever has been doing the munching.[45] It might not sound correct to say that plants "choose" to produce toxic chemicals in response to marauding insects or that water fleas cleverly respond to the presence of predatory fly larvae by growing helmets, but behavioral and physical forms of response differ only in their flexibility and speed of action. One response is calculated by a fast computer built using neural networks and the other by a slower computer that relies on gene regulatory networks, but both involve active computation.

Some environmental influences can produce responses that persist over multiple lifespans, such as the programmed response to malnutrition found in humans and many other animals. People who are nutrient-deprived during their childhood do not grow as tall (on average) as those who are well-fed. Contrary to what is often assumed, the short stature of malnutrition is not a direct result of missing nutrients but reflects a judicious evolutionary strategy based on the premise that smaller organisms can get by with less food. Inadequate nutrition is a signal, and GRNs involved in development recognize and respond to that signal by limiting growth.

Remarkably, this growth-limitation response operates over multiple generations. If my mother is seriously malnourished while I am in utero, I am likely to reach a shorter adult height, even if my own nourishment is adequate. And if it was my mother's mother who suffered malnutrition during her pregnancy, there is still a good chance that I will end up shorter than otherwise.[46] One might speculate that this adaptation evolved because when famines occur, they often persist for multiple generations.

The reduced heights of two subsequent generations result from a type of temporary chemical modification to DNA known as methylation. Because such "epigenetic inheritance" involves no changes to DNA base pairs, it eventually wears off. Epigenetic methylation seldom continues further than two generations, but certain other kinds of epigenetic influences, mostly found in insects, can persist for as long as 50 generations, easily overlapping the range of the more rapid forms of Darwinian evolution.[47]

While these are (perhaps) somewhat unusual examples, organisms with a single genotype typically express a range of adaptive phenotypes in response to environmental conditions. In this essay, I will refer to all the adaptive phenotypes that an organism with a given genotype can express as examples of that organism's *phenotypic plasticity*, where "plasticity" means something like "ability to change."

Caveat: The terminology in this area of biology is something of a minefield. There are multiple definitions of phenotypic plasticity in use, and some make relatively fine distinctions between phenotypically plastic responses and "reaction norms." However, I will follow the recommendation of Mary Jane West-Eberhard and group all forms of adaptive phenotypic response together (West-Eberhard 2003, p. 33).

Adaptive phenotypic plasticity emerges largely under ordinary natural selection on variation operating on Timescale Two: Organisms better able to respond to the range of environments that they are likely to encounter during their lifetimes are more fit than those that cannot. Subtler processes can induce phenotypically plastic responses to intermittent conditions that organisms experience only every few generations.* But no matter how it comes to be, phenotypic plasticity operates as a generalization over the conditions that can be expected to occur at Timescale One.

Adaptive phenotypic plasticity offers multiple layers of response, some of which act as catch-alls or backstops when others fail or are neglected. Suppose I often develop blisters on my feet when I hike. To address the problem, I can use my sophisticated central nervous system and shop for boots with a wider toe box. If I like the way those shoes feel, I might start to buy all my shoes from the wide toe box category. This would be an example of behavioral phenotypic plasticity. However, if I fail to change my behavior and continue to wear chafing boots, slower processes will handle the problem their own way, and I will develop strategically placed calluses.

The development of calluses is not mediated through a brain, but it is still a flexible, adaptive form of phenotypic plasticity, an implicit response pattern that, if converted into an adaptive proposition, would state, "When skin is repeatedly rubbed, build a callus there."

Phenotypic plasticity can allow organisms to imitate evolutionary adaptation on a rapid but more limited scale. For example, goldfish typically attain an adult size in rough proportion to the extent of the body of water they live in, and the size difference between a goldfish in a fishbowl and one in a lake is so great that at first glance, they might appear to belong to different species. (This sliding-scale size adjustment is a model of models representing the many sizes of goldfish bodies that are most likely to do well in bodies of water of various sizes.) Similarly, chimpanzees forced to live in hot, savanna-like conditions in Fongoli,

---

* When the selective environment of a population of organisms changes back and forth over periods longer than a single lifetime but too rapidly for mutations to catch up, mutations that cause reasonably effective plastic responses may be favored over those that offer more effective but narrowly tailored ones. This effect can be thought of as a kind of evolutionary friction that acts against selective processes. For a formal analysis, see Rago A, Kouvaris K, Uller T, Watson R. How adaptive plasticity evolves when selected against. *PLoS Comput Biol.* 2019;15(3):e1006260.

Senegal, have taken to sheltering in caves and hunting with sharpened sticks, almost as if they wanted to follow the same path that converted our common ancestor with chimpanzees into humans.[48]

The near-evolution-scale, guided adaptations produced by phenotypic plasticity come in handy when organisms reach the limits of their capacities and begin to come under directional selection.

Optimization on Timescale One Guides Optimization on Timescale Two via the Baldwin Effect: Phenotype-First, or Genes as Followers

When environments shift, such as during the onset of an ice age, the new conditions challenge whatever forms of optimization arose and stabilized during prior eras of slower change. Organisms initially respond to novel environmental conditions by using all the available resources of their adaptive phenotypic plasticity. Over longer periods, selective forces swing into action, but these necessarily operate on the shifted phenotypes that have emerged under stress rather than on the original, unstressed phenotypes. This two-step process allows phenotypic plasticity to intelligently redirect the loci of genetic evolution toward adaptations that are likely to prove useful. This is the final element of Kirchner and Gerhart's theory of facilitated variation. To illustrate how it works, I will reuse a hypothetical example I presented in my book *Cooperation and the Evolution of Human Nature*.[49]

Suppose that food resources in a river begin to dwindle, but plenty of nutrient-rich prey are available out in the ocean. Because fish are active agents with considerable problem-solving abilities, they will swim out into the river's somewhat salty estuary as far as they can tolerate. Ocean water is unpleasant and, ultimately, toxic for fish adapted to fresh water, but most freshwater fish can tolerate at least short forays into salty conditions. Prior to this change in behavior, selection operated on fish comfortably ensconced in fresh water; now, it operates on genetically unchanged fish that are regularly darting out into salt water in search of food and returning better-fed but somewhat sick from the experience.

This shift in behavior causes selection to "pay attention to" a form of variation that it had previously failed to notice: differences in innate salt tolerance. Those fish that, by chance, possess the greatest ability to withstand saltiness will be able to obtain more food and will do well under selection (will be "selected for"), while less resilient fish will do poorly (will be "selected against").* Over time, GRNs and genes whose phenotypic plasticity can stretch to saltier water will become more prevalent in the fish population through the ordinary processes of natural selection; the altered behavior is said to have been "genetically accommodated." But, as always, most such accommodation probably occurs in non-coding DNA regions rather than in literal genes.

Note that it was the fish population's self-directed behavior that induced directional selection for increased salt tolerance. If our challenged fish hadn't voluntarily pushed themselves to their salt tolerance limit, chance mutations that enhanced salt tolerance would not have come under selection and evolution in that direction would not have occurred. The intelligent, food-seeking processes coded into their adaptive phenotypic plasticity shifted the region of phenotype space that evolutionary adaptation could begin to explore. Because Darwinian selection is blind but organisms are not, phenotypic plasticity can significantly speed up evolution.

For an even simpler example suggested by Watson,[50] consider the archetypal transition from fish to amphibian. Ordinary fish have cartilaginous fins that they use to swim in water; lobe-finned fish possess fins augmented with muscle and bone that they use to drag themselves from one drying pool to a better one; amphibians traverse land surfaces by walking on splayed-out legs. The traditional, genotype-first analysis would have to claim that this transition was led by mutations that made it slightly easier for ordinary fish to move on land. However, this

---

* The astute reader may note that the presence of an inherent ability to withstand saltiness implicates a second form of plasticity at work in this story: the capacity to adjust physiologically to increased saltiness. More on this shortly.

seems to reverse the direction of cause and effect. A more reasonable description would be, "If you are already trying to walk, selection will reward mutations that make you better at walking."

The process of phenotypic plasticity guiding variation in adaptive directions is commonly referred to as the Baldwin effect, named after American psychologist James Mark Baldwin, who wrote presciently about the phenomenon in the late 19th century.[51]

Once genetic accommodation has made it easier for organisms to achieve states that have been trialed by phenotypic plasticity, further modifications may make the change normative; the fish of this story may eventually become natively saltwater fish unable to tolerate fresh water for long. If this happens, genetic accommodation can be said to have matured into genetic assimilation (although this is an oversimplification of both those terms). In the words of the philosopher of life sciences Daniel Dennett,

> Thanks to the Baldwin effect, species can be said to pretest the efficacy of particular different designs by phenotypic (individual) exploration of the space of nearby possibilities. If a particularly winning setting is thereby discovered, this discovery will create a new selection pressure: organisms that are closer in the adaptive landscape to that discovery will have a clear advantage over those more distant.[52]

Evolution led by phenotypic plasticity is summarized in the phrase, "Genes are followers rather than leaders," or, "phenotype-first." 0West-Eberhard is the great exponent of this idea. Her monumental and mind-stretching work *Developmental Plasticity and Evolution* argues that evolution seldom advances through serendipitous mutations in genes or GRNs but almost always follows phenotypically plastic responses that are later accommodated by genetic change (West-Eberhard 2003).

While this claim is contested by some, little doubt remains that phenotypic plasticity is at least a major contributor to evolutionary change.

I cannot possibly do justice to the full range, depth and subtlety of current thinking about phenotypic plasticity, but I will explore a few additional examples to show the close relationship between phenotypic plasticity and the related phenomena of induction, generalization and evolutionary learning. I apologize in advance for many simplifications that will strike experts as over-simplifications; to flesh out these ideas fully would require more background detail than this essay can support. West-Eberhard 2003 is an essential resource.

Behavioral plasticity is not the only form of phenotypic plasticity that implicates the Baldwin effect; physiological and anatomical plasticity can also pave the way for rapid evolutionary change and may commonly play a dominant role.* This element was already present in the hypothetical freshwater fish story because, without the preexisting physiological plasticity that permitted fish to survive brief salt water forays, and their homeostatic ability to detect when they needed to head back to avoid salt-induced injury, the behaviorally plastic response of darting out toward the ocean could not have occurred.

A non-hypothetical example of genetic assimilation led by anatomic plasticity appears in certain physical adaptations accomplished by that famously over-studied fish, the stickleback. Most sticklebacks of the three-spined species live in the ocean. However, toward the end of the last ice age, 15,000 or so years ago, some began to colonize newly appearing freshwater lakes. Those stickleback that ended up in deep lakes came to possess upturned jaws adapted to feeding on floating plankton, while those that found themselves in shallow lakes adopted

---

* The origin and evolutionary effects of developmental phenotypic plasticity are more complex than indicated in my presentation here. For an up-to-date and accessible analysis, see Uller T, Milocco L, Isanta-Navarro J, Cornwallis CK, Feiner N. Twenty years on from Developmental Plasticity and Evolution: middle-range theories and how to test them. *J Exp Biol.* 2024;227(Suppl_1):jeb246375.

a bottom-feeding lifestyle and now possess a mouth and jaw structure more useful for stirring up sediment and capturing invertebrate prey.

The transformation of saltwater stickleback into two new freshwater forms has proceeded much too quickly for random genetic mutations to have accomplished it. Clues about the factors permitting this rapid divergence have been discovered through research on the original three-spined stickleback, which, luckily for interested evolutionary biologists, still lives in the ocean.

When researchers expose marine three-spined sticklebacks to the food sources and other conditions experienced by bottom-feeding freshwater sticklebacks, their mouth structures rapidly adopt the appearance and behavior of the bottom-feeding variant.[53] Conversely, exposure to a diet and conditions resembling those of the deep-lake form initiates a shift toward the deep-lake mouth type. It appears that the progenitor saltwater species possesses an innate within-life ability to modify its mouth shape and structure in response to conditions, and some of these modifications closely resemble the mouth shapes and structures of the two freshwater subspecies. (Within-life plasticity of mouth structures is widespread; it is the reason why sucking one's thumb as a child can cause upper jaw overbite and lower jaw underbite, and why braces can fix it.)

So far, phenotypic plasticity has done all the work for this shape-shifting stickleback. But in a second step, gene/GRN changes followed along and supported the phenotypic adaptations. Just as in the hypothetical example of freshwater fish seeking food in the ocean, selection pressures favored those sticklebacks that happened to be the most capable of morphing their bodies into the appropriate body type. Through multiple rounds of phenotypic plasticity and genetic accommodation, the evolving stickleback will optimize form to function and gradually take on more extreme and finely adapted versions of the original plastic responses. This process has already begun. The two freshwater stickleback varieties are unable (or, at least, less able) to morph into one another than the progenitor species, which can morph into either form. They are becoming genetically and not just phenotypically distinct; their genotype/GRN-type is beginning to fully assimilate one or the other end of their prior phenotypic range.

In general, when an organism can change its behavior, anatomy or physiology to suit conditions A or B, it is primed to evolve into two genetically separate species: one specialized for A and the other for B. When this happens, each offspring species will look a lot like the phenotype variants of their parent species but taken a bit further. Thus, phenotypic plasticity can be the mother of new species. (Further research is needed to determine how often this occurs in real life.)

Similar processes may have played major roles in human evolution, although here we move from observable data to informed speculation. The presentation here follows the ideas of philosopher of life sciences Kim Sterelny.[54]

Most anthropologists believe that climate change was an important initiating factor in the evolutionary divergence of humans from other apes. According to this view, a prolonged cooling and drying spell compelled the ancestors of protohumans to move from dense forests into the open conditions of savannas, much as has happened to the Fongoli chimpanzees mentioned above (although for different reasons). One consequence of living in the open is that predators can spot you from a distance. A former forest ape finding itself so exposed might plausibly use its native intelligence and briefly stand upright to check for lions and hyenas. Chimpanzees and gorillas can stand upright briefly and walk upright for short distances, but doing so tires them out because their body structures are optimized for knuckle-walking. But when environmental conditions became such that a determination to stand for prolonged periods enhanced survival, selection favored the apes *most* capable of standing at will, and, in time, genetic accommodation gave rise to the obligate upright stance of *Homo erectus*.*

Physical, anatomic plasticity would have made this transition easier, as shown in a remarkable story unearthed by West-Eberhard—the true tale of a goat born without functional forelegs.[55] This congenitally disabled creature adapted to its plight by teaching itself to walk and run on its hind legs. When the goat died, anatomical dissection revealed that the bone and muscle structure of its pelvis had diverged considerably from the normal anatomy of goats; the two-legged goat looked like a strange mutant born to stand upright. But none of those changes to the bones were genetic; they were within-life adaptive responses to the goat's effortful upright stance.

As mentioned above, an organism's bone structure is powerfully influenced by the stresses placed on its bones during development. The same process continues after birth as a form of phenotypic plasticity. Muscles, too, respond to the tasks they are asked to perform, and also to changes in bone structure. The two-legged goat's body had partly adapted to an upright stance because, from shortly after birth, it had to stand upright if it wanted to move around, and physical adjustments followed as a result.

According to this scenario, our ancestors found themselves in an analogous situation on the savanna. There are (happily) no stories of apes born without functional arms, but if it were to happen, we might find that the ape's pelvic and bone structures would come to look rather like our own. Genetic accommodation might not have had to fiddle around too much to establish the body type of *Homo erectus*.

The power of phenotypic plasticity to produce species change is enhanced by the fact that environmental influences act on every member of a population at once and will induce responsive phenotypically plastic

---

* Standing upright to spot predators is not the only proposed explanation for the emergence of an upright stance, but all theories begin with voluntary upright standing that is then genetically assimilated.

changes in many or most of them; these changes can then undergo accommodation and assimilation. Mutations, by contrast, occur only in individual organisms that must still evade many other possible causes of death or, at least, failure to reproduce before they can come to dominate a population. Furthermore, while phenotypically plastic changes can markedly increase fitness on the spot, most mutations only slightly enhance fitness when they first emerge and require their own kind of genetic accommodation—a series of supportive mutations—before they provide much benefit. For these reasons, phenotype-first evolution is typically more efficient than one led by mutation.

In addition, adaptive phenotypically plastic responses are *intelligent* in a way that blind variation and selection are not. This intelligence is most apparent in the case of behavioral phenotypic plasticity implemented by brains, but numerous other processes are also intelligent in the sense that they "choose" appropriate responses to environmental challenges. Examples discussed above include increasing bone density and size in response to physical stress, growing a larger head in the presence of predatory fly larv and creating toxic chemicals to keep herbivores away. The intelligence built into phenotypically plastic responses takes evolution by the hand and guides it through traffic. If freshwater fish hadn't been clever enough to deliberately seek out prey in salt water, they might have had to adapt in other, plausibly less effective ways, such as becoming smaller so that they didn't need as much food; nonetheless, that, too, would have been an intelligent response.

As mentioned above, the process of evolutionary optimization is called "climbing a fitness peak," where the peak represents the fittest state that can be achieved through variation and selection. Computer algorithms that climb fitness peaks—hill-climbing algorithms—tend to get stuck on hillocks or knobs, the many small bumps far short of a mountain's true peak; evolutionary processes and (almost) all other forms of optimization run into the same difficulty. This is a primary way the evolutionary past can constrain future evolution: Once on a local fitness peak, evolution cannot easily climb down again, even in the service of subsequently going further uphill, because selection blocks changes that reduce fitness. Evolving organisms can't borrow from future fitness; each step must increase fitness—or, at least, not harm it.

Much the same problem often stifles political innovation; for example, revamping a healthcare system might ultimately benefit everyone (reach a better fitness peak), but the short-term pain of significant change from the current system (a local fitness peak) can put this out of reach. Thus, most healthcare reform plans, like other forms of social reform, typically proceed by attempting to further optimize the current system rather than jumping to one that is entirely different, even if better.

Nonetheless, evolving organisms do not necessarily become trapped on fitness peaks forever. One possible way they can escape their predicament leans on the multidimensional character of evolutionary adaptation: An evolving organism may become trapped on a fitness peak for one trait, but once having reached that peak, it may become able to hill-climb up the gradient that optimizes a different trait. Each new adaptation changes the fitness landscape, or at least brings organisms to a new place within it, and may present new opportunities.

Alternatively, environmental conditions may shift in ways that flatten a previously climbed hill; returning to the human example of healthcare systems, the general destruction of European societies after the Second World War made it possible for them to design and build relatively optimal healthcare systems from the bottom up.

But phenotypic plasticity offers a different way out, one that resembles dynamic leadership. Organisms can use their various evolved capacities to identify distant fitness peaks and move toward them. For the freshwater fish of our story, that new peak was life in the ocean, and their instinct to search for food, their tolerance of salty water, and their ability to know when it was time to head back directed and permitted them to test it out. Phenotypically plastic capacities can identify a distant fitness peak and carry current organisms up to its base; next, ordinary mutational hill-climbing processes in the form of genetic accommodation take successor organisms to the top. In this way, intelligent phenotypic optimization processes enhance the "guess" half of evolution's guess-and-check algorithm by setting it going in the right direction. In principle, blind selection could find the same solution, but perhaps too slowly to succeed in real life; after all, if an entire population dies because it fails to adapt in time, no mutational variations can resurrect them. Phenotypic plasticity is much faster.

It is also potentially safer. An organism can observe the near failure of a behavioral strategy and subsequently modify it, whereas genetic changes are more difficult to undo. Social animals can use an even safer strategy: observe the results of behavioral strategies trialed by others and copy or avoid them depending on how they work out. Rats are famously good at this, but so are many other social animals.

The Baldwin effect is an example of self-action in evolution. Phenotypically plastic responses acquire their adaptive intelligence through evolutionary processes, but they then turn around and guide future evolution; by doing so repeatedly, they iteratively inject intelligent decision-making into the blind processes of evolution. To borrow a term from computer science, intelligent phenotypic responses loosely resemble *oracles,* sources of reliable information that guide the decision-making processes of a system and permit it to limit the use of inefficient trial-and-error. The oracles provided by phenotypic plasticity are far from infallible, but they do not need to be; they simply need to perform one percent better than chance.*

And, as it should go without saying by now, phenotypic plasticity facilitates evolutionary adaptation to novel environmental stresses despite having never come under selection for its capacity to speed up evolutionary adaptation. Plastic responses emerge through the slow processes of adaptive evolution because they offer benefits over a single lifetime (or, perhaps, a few generations). However, just as with the evolution-enhancing effects of skeleton-key widgets, the capacity of

---

* The same low floor (and often ceiling) applies to human experts, political pundits and economic models.

phenotypic plasticity to facilitate future evolution is not particularly mysterious. This power lies in the extensibility of models beyond the data upon which they were founded.

Phenotypic plasticity models the range of Earth environments an organism is reasonably likely to encounter as it goes about its daily life and offers effective responses to them. However, the environmental shifts that can occur during an organism's lifetime typically sample those that occur on longer timescales. When an environment shifts, prior extremes may become norms, but the occasional experience of those extremes, or changes in the same direction, will often have prepared an at least partially effective adaptive response. And even when conditions go past anything previously experienced, existing phenotypic plasticity can often stretch beyond the range of the environments for which it evolved; scientists have never, but really should have, called this process the *Spinal Tap effect*—turning an amplifier originally designed to go up only to 10 further up to 11.[†] Freshwater fish can briefly push themselves fairly far into salt water; chimpanzees can stand upright, even if it rapidly exhausts them; and humans, whose brains must have been fully modern 100,000 years ago (and probably sooner), subsequently used those brains to invent calculus and go to the Moon.

West-Eberhard goes further along the same lines and persuasively argues that *most* evolutionary adaptation is preceded by and goes along the lines of phenotypic plasticity. A population of worms with eye spots might randomly develop the precursors to a camera eye, but incremental changes in that direction are unlikely to provide selective benefits unless these worms already have a vision-centered lifestyle; only those creatures currently doing everything they can to squeeze the

---

† Undoubtedly, the only reason biologists have refrained from using this expression is that they are extremely precise in their use of pop culture references, and the original joke in *This Is Spinal Tap* made a different point: The lead guitarist foolishly believed that an amplifier whose volume dial reached to the number 11 could necessarily get louder than one whose dial stopped at 10. However, in cultural memory, the expression has morphed to include the idea of exceeding the original design specifications of a system.

maximum bits of information out of their existing visual systems will undergo directional selection for improved vision. The ghostly outline of a camera eye already exists in a worm's striving to see.

Phenotypic plasticity is the leading edge, the toe in the water of evolutionary change; it doesn't just facilitate variation but actively and intelligently provokes it into being.

Optimization Timescale Three: Hebbian Relationships and Modularity

We have now gone through all the primary elements of facilitated variation considered by Kirschner and Gerhart. In the final third of this essay, we turn to the learning systems approach to evolution developed by Watson and his research associates. This novel conceptual framework makes at least three significant contributions to the theory of facilitated variation.

Most simply, it offers a natural and intuitively transparent explanation of the origin and persistence of pluggable modularity.

Somewhat more complex but still intuitively plausible arguments based on this framework, together with formal computer simulations, suggest that evolutionary processes should tend to retain "memories" of past adaptations and the relationships between them. When environmental conditions occur under which they again enhance fitness, those adaptations can be expeditiously resurrected, even if they have long fallen out of use and become partially corrupted.

The third consequence of their analysis is more radical, less intuitively obvious and potentially the most profound.

As the GRNs of organisms evolve over Timescale Three, they acquire patterns of independent and linked traits that represent regularities in Earth conditions. This is a modeling process that formally resembles the sequential transformations that occur in brains as they learn about the world and in artificial neural networks as they train on datasets. Because of this close relationship, when directionless mutations shake up the developmental machine, the deep GRN structure of that machinery in effect generalizes from its past discoveries and elevates the probability that those mutations will produce good guesses—adaptive variations.

This is the theory that I have been calling natural induction through self-modeling. If proven true, it will go a long way toward explaining the origin of facilitated variation and may also provide answers to other long-standing evolutionary riddles.

Watson's ideas are somewhat challenging, and I will approach them gradually. Actually, I have already begun to do so. My many references to induction, generalization, optimization, modeling and neural network learning above were intended to familiarize the reader with ideas that will be important in the following three sections. It's finally time to dig in.

Evolution engages in a form of slow computation as it transforms one organism into the next. If we regard the total set of GRNs contained within an organism's DNA as a data structure, subsequent Darwinian selection on that organism's descendants produces a series of data structures, each slightly different from the one that came before. In certain important ways, these sequential GRNs resemble the sequential structures that emerge as a neural network learns to identify images of cats or builds a model of a human language; to paraphrase Watson, GRNs evolve very much like neural networks learn (Watson et al. 2014).

When a large language model trains on text, its data structure acquires values and relationships that model those texts. The model that emerges can predict what words are likely to follow other words, what sentences should come after preceding sentences and how paragraphs should follow paragraphs in a coherent longer text.* If a well-developed large language model is given a sample of novel text, it can edit,

---

* To be precise, the system predicts individual words, but because word prediction takes into account the structure of surrounding words and punctuation, it, in effect, operates on higher-level textual structures too.

continue, improve or respond to it with a high probability of doing so competently.

According to Watson's paradigm, the sequential DNA data structures produced by evolution similarly reflect ongoing training, in this case, on what adaptations permit organisms to survive and thrive as they reproduce in varying environments. Over Timescale Three, these data structures have become a model of *how* to produce appropriate, responsive adaptations by combining sub-adaptations and sub-sub-adaptations. When organisms come under selection in novel environments, the GRN structures that control development guide the effects of directionless genetic mutations toward phenotypic variations that have an elevated probability of proving adaptive in the new conditions; metaphorically speaking, genotypic mutations are shepherded by GRNs into phenotypically educated guesses.

All three of these ideas flow from a singular insight that links GRN modularity to associative learning in neural networks.

What Watson Recognized

Evolutionary biologists have known since Darwin that some traits evolve independently while others are tightly linked. A mutation can alter the color of a bird's tail feathers without changing the shape of its beak, but any genetic shift that alters the shape of its right wing will also affect the left. Covarying features co-evolve, and independent features evolve independently.

In the 1990s, evolutionary biologist Günter P. Wagner noted that traits that commonly come under selection together become developmentally linked.[56] If there is a fitness benefit for organisms to possess both traits A and B, and that benefit exceeds the fitness benefits (if any) offered by A or B alone, one or more developmental processes will evolve to help ensure that the two traits arise together. Similarly, when traits C and D have an optimal relationship to one another, such as the 1:1 ratio of front and back leg length in quadrupeds, processes to fix that ratio in place will tend to evolve. If a descendant organism living under similar conditions undergoes a mutation that interferes with the embryonic development of trait A, the presence of trait B can restore trait A through one or another backup system; if traits C and D are meant to arise in a fixed ratio but they begin to drift away from that optimum, similar backchannel connections can restore the optimum ratio.

Watson seems to have been one of the first to recognize that this linkage process functionally resembles the fundamental mode of operation of animal brains: associative learning.[57]

Associative learning is a familiar process. Anyone who has raised a cat or a dog knows that house pets rapidly come to connect the sound of opening a bag of dry food with the appearance of food in a bowl. Rattling the bag gets them just as excited as letting them smell the food. The process of building associations is so fundamental to the operation of nervous systems that its operation can be detected in the physical operation of neurons: Neurons that typically fire independently remain independent, but those that frequently fire at the same time or in close sequence acquire a strong synaptic linkage. This phenomenon is called Hebbian learning, named in honor of Donald Hebb, who first identified it in 1949.[58] Ever since, students of neuroscience have learned the mnemonic phrase "neurons that fire together, wire together." If one of the two connected neurons is stimulated to fire by outside events, the synaptic connection between them will cause the other to fire, too. This is why pets respond to the sound of a rustling food package in much the same way as to the sight and smell of the food within it.

The evolutionary processes that construct redundant linkages between associated traits also operate as a form of Hebbian learning, but one that connects traits or adaptations rather than neurons. Instead of "neurons that fire together, wire together," this is "traits that go well together evolve to develop together." (A prize to anyone who invents a mnemonic that rhymes.) To keep this similarity in mind, I will from this point on expand the use of the term "Hebbian" to include linked traits as well as linked neurons.

To understand the implications of Hebbian trait linkage, let's first review how associative learning operates in the field where it was first observed: biological neural networks.

Associative Learning

Hebbian or associative learning is the basis of Pavlovian conditioning. If a dog repeatedly hears a bell before being presented with food, it will soon begin to salivate after the bell rings, even if no food follows. The dog's brain circuits that trigger salivation have become connected with those that identify the sound of a bell ringing. They were not wired together at birth but become wired together through experience. This can be described as a form of self-modeling in which repeated patterns of neural activity induced by outside events become internalized into neuronal structures. Nervous systems thus learn to model their own past behavior (and, in doing so, the world). This has many consequences.

While Pavlov's example involved one event preceding another, association also occurs for simultaneous events as well as events in reverse temporal order. Nor are multiple repetitions always necessary. My daughter once happened to experience symptoms of a stomach virus shortly after eating a pomegranate. For many years after, she felt nauseated at the mere sight of the fruit. Once a connection of this kind is forged, it tends to persist, and if an associative memory begins to fade, a single reminder may be enough to restore it.

In addition to correlation, associative learning also identifies events and features that are anticorrelated. If I put on my coat near bedtime to get something out of my car, my dog doesn't get excited because she's learned from experience that I never take her out for a walk once I've turned off all but my bedside lamp in preparation for going to sleep. The state of "lights out" is strongly correlated with staying inside and hence anti-correlated with the state of "going for a walk."

Mathematically, these are all forms of correlation; anti-correlation is correlation with a negative number. However, to keep the narrative non-mathematical and intuitively simple, I will largely neglect anti-correlation in what follows; the same arguments work with a few words reversed.

The neural networks that constitute brains are largely constructed of Hebbian associations that link temporally correlated events, perceptions, features, actions and other elements of significance. Importantly, Hebbian association is compositional: Groups of associated items can themselves become associated. I associate my rare trips to fancy restaurants with the combined experiences of a glass of wine, an appetizer, the main course and a dessert; I further associate "wine" with a set of drinks and "dessert" with a set of sweet foods. Composition causes neural network Hebbian associations to develop nested, hierarchical structures of relationships.

In general, we perceive and understand the world by decomposing it into associated features composed of associated features stacked many levels down. For a well-studied example, consider how the human visual system has evolved to recognize objects of interest. The retina is composed of rods and cones that resemble the pixels of a digital sensor, but the brain doesn't receive the raw data. Instead, the cell layers immediately beneath the retina process incoming images the moment they are received, and it is this processed data that reaches higher brain centers. Initial processing decomposes visual input into such features as edges, shapes, colors and motion. Subsequent layers deeper in the brain's visual processing systems catalog how these features relate to one another and, even further in, how those relationships relate to one another. We learn to recognize objects not by memorizing their appearance bit by bit but by learning the relationships that characterize them. To oversimplify, in the image of a cat, many lines together compose fur; a circle with a vertical slit inside it is an eye; and a face is a roughly circular object with two eyes, a nose and a mouth—each of which is composed of sub-features and sub-sub-features.

One can draw a meaningful analogy between how biological and natural neural network systems recognize cats and the way that developmental systems build them: Both decompose a whole into features composed of features composed of features many levels down. The modular GRNs that build a cat through embryonic development construct anatomic features composed of sets of features that are themselves composed of sets of features—in this case, not of external appearances but of every living part.

The method of identifying objects by decomposing them into visual features and sub-features reduces cognitive load. Even people described as capable of total recall cannot remember or even perceive everything that strikes their retina because that raw data is immediately processed; at most, they accurately recall a predigested image broken into the features our visual processing systems evolved to identify and recognize. The specific way that we decompose images into features itself contains information and is a product of evolutionary learning; it uses techniques of visual analysis that trial and error has found to be useful for survival. GRNs similarly decompose environmental conditions into responsive adaptations built of sub-adaptations, grouping and dividing them according to a system that has been progressively refined over hundreds of millions of years.

The process of visual analysis by decomposition into features continues up through more recently evolved brain centers, although along the way, it becomes progressively less hardwired and more subject to within-life learning. When a mouse scans its environment, it mostly sees food, threats and hiding places, perhaps a cat, a piece of cheese and a hole in the wall. Its brain models elements of the world that fall into one of these classes as (more or less) interchangeable representatives of that class.

Classes are sets of relationships built out of sets of relationships built out of sets of relationships; they are chosen because they bring together functionally similar things. To a mouse, a large gray cat perched on a washing machine and a small tabby hiding behind a sofa lie close together in "mental representation space." However, a perfect photograph of the same large gray cat perched on a washing machine lies far away from the living model in that space despite its similar appearance. A dead cat, too, is not a cat even if it looks exactly like one; the visual categorization system that mice rely upon to thrive separates living cats from dead cats and real cats from photographs of them.

When faced with an unknown image, visual representation systems classify its elements and, in effect, ask, "Is this something irrelevant, like a shoe or a piece of paper, or is it a mousehole, a piece of cheese or a cat?" If the best match is "cat," those systems ring a loud bell of alarm.

This method of bringing like images together and keeping unlike images apart evolved because it is crucial for survival. Mice that can only identify cats they have seen before will soon be eaten; nor would it be helpful to a mouse if it ignored Tabby-the-Excellent-Mouser because it saw her in profile rather than face-forward or in broken shade rather than bright light. Generalizing across a broad set of visual representations to yield the single category "Cat ahoy!" is useful because, from the perspective of a mouse, one cat is much like another, and each can be regarded as a separate instance of a single thing. No matter whether it is a black cat or a tiger-striped one, the best response is to run for the nearest mousehole.

The detailed Hebbian relationships encoded into developmental systems also bring like things together and keep unlike things apart. Just as living cats of widely varying appearance lie near one another in visual representation space, plausible phenotypic adaptations lie close in mutation space to the current phenotype; the mutational distance between a small deer and a larger one is small and likely to occur by chance (although in stages). Conversely, the mutational distance between a deer with normal legs and one with an elongated right foreleg is large and unlikely to occur, because it is unlikely to prove adaptive. Nearby phenotypic alterations offer adaptations responsive to ways that Earth environments have shifted in the past, such as increased predator size. When environmental conditions do change, certain of those phenotypic variations are likely to prove adaptive and come under positive selection. The emergence and selection of an appropriate suite of responsive adaptations functionally categorizes the current environment as a structurally similar variation of one experienced in an organism's evolutionary past and provides an appropriate adaptive response. This mechanism improves the "guess" half of evolution's game of guess and check.

A more detailed exploration of these challenging ideas follows.

How Developmental Processes Build Hebbian Relationships

For an embryonic developmental system to construct almost any useful adaptation, multiple features constructed by multiple GRNs must be brought together. Bones aren't too helpful without joints and joints make no sense at all without bones. To produce a composite structure of bones and joints, the developmental processes that construct these structures must be activated at the same time or at least in close sequence during embryonic development. When developmental systems emerge that link related structures, the equivalent of a Hebbian neural relationship is born.

A simple version of Hebbian trait linkage occurs in bacteria: The genes that code for functionally interdependent proteins are arranged contiguously as an operon. This arrangement permits a single ON/OFF switch to control their simultaneous expression. Eukaryotes use more flexible processes. One common design motif activates an identical module in more than one embryonic compartment. This is how bilaterally symmetric animals maintain their symmetry: The modules that build one side of the body are also used for the other.

Alternatively, eukaryotes may instantiate Hebbian linkages by adjusting the settings on the control boxes that activate multiple different processes in such a way that they all respond to the same incoming transcription factor. For example, Pax-6 might bind to the control boxes that activate both lens- and retina-forming processes. A different method might run the linkage sequentially: Pax-6 initiates retina formation and then the retina itself goes on to initiate the lens. Or, if evolution didn't find that approach convenient, it could arrange matters so that Pax-6 simultaneously triggers the release of two different transcription factors, one of which activates lens formation and the other retina formation. A bit of imagination will suggest many other possible linkage mechanisms.

No matter how they are implemented, Hebbian links between traits emerge under selection. Here is a schematic example to show one way it can happen:

The right and left pectoral fins of fish are adaptations for swimming. With rare exceptions, fish get on best when the lengths of their paired fins are closely matched. Suppose (undoubtedly contrary to the facts) that when right and left fins first evolved, their lengths varied independently. Due to the independence of these traits, only a small number of fish would possess relatively well-matched fins at birth. However, fish with somewhat matched fins have a strong selective advantage over those with badly mismatched fins; less-balanced fingerlings tend to be eaten. Each current population of fish will, thus, lose more of its less-symmetrical members to predation, and the surviving population will consist of fish with fins that are less badly mismatched. Nonetheless, since the traits of "right fin length" and "left fin length" are still independent, each subsequent generation will start over with whatever random fin-length relationships happen to occur; the observed fin length semi-equality of surviving fish is a result of ongoing predation rather than a genetic predisposition. Fin lengths are *correlated* in fish populations, but there is as yet no *causal* developmental process to produce matched fins.

But one will soon emerge.

Suppose that random GRN variations cause some fish to possess heritable developmental processes that are slightly biased toward producing roughly-matched fins. The effect doesn't have to be strong; perhaps these processes reduce the average difference in fin length by only a few percent. Nonetheless, because fish with more nearly-matched fin lengths are more likely to survive, GRN modifications that mildly enhance fin-matching will become increasingly common in the population. Correlation in populations due to fitness differentials has been transformed into causal developmental pathways. This process will iterate until all fish are born with quite closely matched fins, the most fit arrangement.

One way to produce better-matched fins might be to start them growing at the same time. Simultaneous initiation could be instantiated in embryonic development by tying the onset of growth on both sides to a single signal that each receives. This is undoubtedly the primary actual mechanism that ensures fin equality; instead of using a separate developmental process for each fin, embryonic development in fish activates the same process twice, once on the right and once on the left, and sets both of them going with a single starter pistol.

Nonetheless, all biological processes are subject to interference. Even if fin growth is simultaneously triggered on the right and left, the two fins might grow at slightly different speeds and continue growing for slightly different lengths of time. If there is a selective advantage to extremely closely matched fin lengths rather than roughly similar ones, the hill-climbing processes of evolution will likely find a way to solve these issues.*

One possible method for improving fin matching would be to employ mechanisms that create robust resistance against the disturbances that might alter fin growth rates. Alternatively, the differential forces exerted by water on fins of different lengths could be used to affect post-hatching growth in such a way that the sizes converge. No matter what mechanisms they use, systems will eventually emerge that ensure a close match between right and left fins.

Through mechanisms like these, whenever two or more features or traits are consistently correlated in populations of an organism because differential selection favors that correlation, those traits will eventually become causally and redundantly correlated.

---

* It should not be assumed that evolutionary processes can always reach any given fitness peak. Some peaks have no plausible path to achieve them, and even those that could be reached in principle may remain out of reach because evolutionary hill-climbing processes get stuck on less optimal local peaks.

Just as with the Hebbian associations between neurons, the emergence of associations between traits is a form of self-modeling, in that selective favoring of animals that accidentally possess more nearly matched fin lengths has been converted to internal developmental patterns that causally mirror those external selective processes. As this process iterates, it gradually converges on limbs of consistently equal lengths and remains there. Settled linkages between traits cause developmental systems to model and "deliberately" produce trait patterns that have recurrently emerged under selection due to consistent patterns in the environment. This is how brains learn and when GRNs do the same thing, they, too, are learning (although much more slowly).

Note that a developmental linkage of this kind can only come under selection if it is *true*; for a Hebbian trait association to emerge in GRNs, feature correlation must provide an average fitness advantage compared to feature independence. Traits whose correlation provides benefits in a large number of circumstances will have that correlation come under positive selection in more than one way and in a greater variety of environments than those where the benefits are sporadic and limited. More closely matched fins help fish in many selective conditions, whether they must chase, dodge, flee and explore in water that is warm, cold, turbulent or still. Because persistently correlated selection is the source of Hebbian trait linkage, the strongest linkages will tend to occur between features that work well together in a great variety of conditions.

These well-established linkages function as inductive generalizations. If converted into an explicit proposition, developmentally enforced fin symmetry states, "Because more closely matched fin lengths have been optimal in many past environments, they are likely to be optimal in most future ones." Of course, this generalization might prove incorrect if a population of fish were transported to some peculiar environment where unbalanced fins work better, such as a circular aquarium, and allowed to evolve there. Because Earth holds few such environments, the generalization holds.

The achievement of perfect matching is itself a form of induction based on following a gradient. Rather than duplicating organisms that were accidentally born with perfectly matched fins, induction duplicates the *direction* of change that enhances fitness. The implicit proposition is "Because fish accidentally born with better-matched fins are more fit, developmental processes that enforce matching will increase average fitness." This remains true until matching becomes good enough that no further selective advantage can be achieved by making the match even tighter.

In a meaningful sense, strong Hebbian relationships represent regularities of the world. Bilateral fin or limb symmetry reflects Newton's third law, that for every action, there is an equal but opposite reaction; more specifically, when attempting to follow a linear path in a liquid or on land, it is most efficient to exert equal forces on both sides of the body. The developmental system for building bilaterally symmetric animals may have originally come into being, at least in part, because it tracks this law of nature.

The example of asymmetrical fins becoming symmetrical is unrealistic because bilateral symmetry is fundamental to all but a few phyla of animals. A more plausible example is the matching of front and back leg length in quadrupeds. Quadrupeds possess front and back legs of equal length because, on average, that is the most efficient design—the world contains exactly as many uphill as downhill passages. Achieving matched leg length would have been an evolutionary challenge to the earliest land animals because the fish from which they evolved did not come under the same selective force; front and back fins need not match because water has no internal surfaces. The linkage between front and back limb length subsequently emerged through processes similar to those described above for bilateral symmetry, and it has largely persisted ever since; however, as in the case of our own ancestors, it has occasionally been discarded, too.

The action of redundant Hebbian linkages reduces the probability that mutations will produce variations (like mismatched limbs) that are likely to turn out badly and increases the probability of variations that evolutionary "experience" suggests stand a good chance of being adaptive. As more "knowledge" of this kind becomes embedded in developmental systems, evolutionary variation becomes constrained to try sensible body plans and processes and to avoid stupid ones.

Traits can also evolve as independent variables. In some cases, they may have emerged independently and never became linked; for example, the developmental systems that build eyeballs and those that build toenails have probably never been closely connected. Alternatively, two traits may have been linked for many millions of years, but when novel conditions occur that are structurally dissimilar to past ones, their linkage may cease to be fitness-enhancing. At that point, any mutation that allows them to diverge adaptively, even to the slightest degree, will come under selection and finally expand through hill-climbing processes until the linkage completely breaks. Tyrannosaurs and other animals that evolved distinct uses for fore- and hind legs have accomplished this.* The divergence between arms and legs would have begun in those organisms whose developmental systems that control front-back symmetry had lost most of their redundant linkages. To call back to the imagery used in the section titled *Switches and Dials*, these were the organisms whose GRNs for trait linkage did not reside square in the middle of the watershed for linked front and back legs but were perched near a ridge, on the other side of which the linkage was broken.

Sometimes, GRN developmental processes are so strongly linked that it takes an event called "gene duplication" to separate them. Various mechanisms can cause an entire section of DNA to acquire a twin. Initially, both versions of the same DNA continue to operate simultaneously, merely replicating one another's function; the quantity of gene products may be problematically doubled, but other systems can come online and tamp down the excess. However, once there are two

---

* As usual, phenotypic plasticity acts as the leading edge of genetic change. Animals must first *try* to use their front and back legs in distinct ways before selective processes can begin to favor genetic changes that de-link them.

distinct sets of DNA sequences that produce a single trait, the second can diverge while the first continues to provide its original functions. For example, quadrupeds presumably use a single patterning module for both front and back legs, but if that module develops a doppelganger, front legs can evolve to use one version and back legs the other, and they could go on to evolve separately.

Thus, over Timescale Three, traits are pushed to be correlated or independent according to the correlation or independence of the environmental conditions to which they are responsive. These trait relationships increase the likelihood that future mutation-induced variation will offer adaptive changes. Most simply, they do so by reducing the rate of "stupid" mutations, such as those that decouple right and left fin lengths or produce a single elongated leg. However, they also facilitate useful variation in more subtle ways. Let's begin to consider them.

Hebbian Relationships Model the Correlational Structure of the World

I once set about trying to teach my young granddaughter the meaning of the word "muggy." At first, she thought that "muggy" and "hot" were synonyms. However, when a dry, hot day finally came around, I took the opportunity to point out that currently, it wasn't muggy, just hot. It took her only that single example to learn the concept; in some efficient fashion, her brain compared temperature to sweat quantity, identified the specific characteristics of damp, hot air that distinguish them from dry, hot air, and tagged it as "mugginess."

Later, she had the opportunity to experience dry, cold air and damp, cold air. Mugginess, she came to understand, is a combination of experiential dampness and heat, and each can vary separately. Other Earth features, however, typically vary together. No matter what the temperature, it's warmer in the sun than in the shade, and it usually feels cooler when the wind blows than when the air is still.

As organisms evolve under varying Earth conditions, some features of their experienced environment vary together while others vary separately.

This patterned switching "teaches" evolving organisms which features of the world are independent and which are correlated (or anti-correlated). The Hebbian GRN structure that evolves in response to these varying conditions thus reflects the underlying correlational structure of the world itself. Adaptations to environmental conditions that vary separately remain or become independent, while adaptations to linked conditions remain or become linked. Some sets of conditions that first appear as unitary may later reveal themselves to be decomposable into separate parts, much as when subsequent experiences taught my granddaughter that the experience of humidity is independent of temperature, and GRN structures will come to incorporate that decomposition.

Thus, over Timescale Three, the Hebbian structure of adaptations imprinted into organisms' developmental systems becomes an increasingly refined, if indirect, model of the world's correlational structure. Because the variations that GRNs are likely to produce for selective trial are constrained to follow that structure, they are more likely to produce adaptive outputs.

The Origin of Modularity in Hierarchical Hebbian Relationships

The above discussion largely focused on trait relationships that occur in pairs. However, just as with associative learning in neural networks, trait linkage is compositional: Sets of Hebbian linkages can themselves become linked. These include both vertical and horizontal arrangements.

To construct an arm, vertebrates build two bones AND a joint between them. The humerus, radius and ulna, together with the joints that connect them, form a horizontal set of Hebbian-linked traits. But each of these linked traits is itself composed of linked traits. For a joint to function properly, linked processes must line the terminal edges of the bones with cartilage, fill the space with liquid and hold the assembly together with ligaments. To build a bone, the process of initial cartilage formation must be linked to another process that attracts bone-forming cells. Thus, each of these subsidiary processes is itself built on linked sub-subsidiary processes that occur at still lower and eventually molecular levels, with branches occurring at each step downward to the level of genes. The result is a vertical hierarchy that controls multiple levels of horizontally linked traits.

A lower-level process is seldom "owned" by a single higher-level process; rather, it may also be activated on other occasions and in other places by altogether different higher-level processes. The subsystems that build the bones of the arm are also utilized to build the bones of the toes, and most genes are probably involved in hundreds or thousands of distinct processes. However, independently varying traits must be produced by processes that can operate largely independently of one another. Thus, the hierarchical structure of the processes that create traits includes sets of traits with complex and redundant internal linkages that are functionally isolated from other such internally linked sets.

We may as well call those sets modules. The fact that multiple higher-level modules can use the same lower-level modules illustrates the "pluggable" aspect of weak linkage; the possibility that previously linked traits can later become independent represents the reverse process of unplugging.

In other words, selection for trait independence and trait linkage produces pluggable modularity for free.*

The same process also *maintains* modularity. A mutation might enhance immediate fitness by reaching across modular lines to directly activate a component of another module, much as if a computer programmer were to write in a hack that breaks software design principles but happens to work. Non-modular hacks in GRN circuitry are "messy" because when they activate one module, they also activate (or suppress) all the other modules that are connected to it through Hebbian relationships. Like software hacks, such solutions may work well under

---

* Does this argument explain *all* elements of biological modularity, including the use of repeated segments and distinct life cycle stages? Maybe, but the question rapidly becomes difficult.

the circumstances where they were first used but fail in others. If the genetic hack fails to prove adaptive in multiple successive environmental conditions, organisms that possess it will frequently lack fitness and the linkage will be selected against. However, if the novel cross-module link proves consistently successful, it will develop, expand and become a new modular element.

Thus, the relationship between Hebbian linkages and modularity is almost obvious once one begins to reflect on it: The modularity of organisms is created and maintained by regularities in Earth environments and in the adaptations that correspond to them.

Much the same idea can be found in the writing of Günter Wagner, Mary Jane West-Eberhard and others. However, Watson and his group, and systems biologist Uri Alon before them, have made the idea more rigorous by using various methodologies to mathematically represent such biological features as embryonic development, traits, environments and fitness interactions. Computer simulations based on those mathematical models suggest that modularity emerges spontaneously under natural selection provided that two factors hold: Environmental conditions must periodically change, and the correlational structure of those conditions changes more slowly than the conditions themselves.[59] This would seem to characterize the actual world, in which, for example, average temperatures vary, and temperature and humidity vary independently, but it is (almost) always warmer in the day than the night and in summer than in winter.

In a deep sense, the emergence of biological modularity resembles the method of solving big problems by decomposing them into subsidiary ones. Why our universe is constructed in such a way that such decomposition often works is an impossibly profound question. Nonetheless, it certainly does often work, and decomposability is exploited as much by the GRNs that manage embryonic development as by brains.

Optimization on Timescale Three Facilitates Optimization on Timescale Two Through Developmental/Evolutionary Memory

The previous section demonstrated the first and easiest to understand of the three claims made by Watson and his colleagues: Hebbian associations are the source (or, at least, an important source) of biological modularity. Now we turn to their second, somewhat more subtle claim: that Hebbian associations create evolutionary memory.

Adaptations develop over Timescale Two, but Hebbian relationships between adaptations arise much more slowly, on Timescale Three. This persistence produces a form of memory for past adaptations that allows modules no longer in use to be restored with relatively few mutations. Since Hebbian relationships emerge only when traits work well together in many conditions, it is reasonably likely that the same traits and trait relationships may prove useful again in the future. Thus, when a population of organisms faces new problems, some of its members may be able to solve them by using tools filed away in the library of past adaptations. This represents a potential form of facilitated variation not considered by Kirschner and Gerhart.

In this section, I attempt an intuitive rendition of a 2014 paper by Watson and his colleagues titled "The Evolution of Phenotypic Correlations and 'Developmental Memory." (Watson et al., 2014). This paper reports the results of computer simulations that follow the evolutionary progression of a population of virtual organisms whose traits and the Hebbian relationships between them are modeled mathematically. Among other results, these simulations suggest that adaptations formed from linked traits will tend to persist long after they have ceased to be used, and even when disuse has allowed adaptations to partially deteriorate, spontaneous hill-climbing processes can spontaneously restore them.

The cited paper uses the term "developmental memory" because the most obvious restoration of traits occurs during embryonic development when modular GRNs produce modular features. However, physiological processes that operate during adulthood are equally capable of being restored after having been abandoned. Since there is no generally accepted expression for the entire set of processes that bring phenotypes into being within a life cycle, I will refer to all forms of recall of adaptations that were once operational in the prior evolutionary history of an organism as examples of *developmental/evolutionary memory*.

I must note from the outset that the model used in these experiments employs a significant simplifying assumption: It only represents pairwise Hebbian relationships between idealized "traits" and does not attempt to simulate hierarchical modularity. Nonetheless, it seems plausible that if even a flat set of Hebbian relationships—one that more closely resembles those found in bacteria than in eukaryotes—can exhibit memory, more complex relationships should do so, too, but even more so.

A simple form of the persistence of developmental/evolutionary memory is readily apparent in the form of atavisms, the spontaneous re-emergence of ancient traits in current organisms. One famous atavism, the appearance of vestigial tails in humans, was discussed above as an example of a developmental switch. We can learn a bit more by studying several other examples.

Whales, Frogs and Snakes

In one famous case, a humpback whale was born with two partially formed hind limbs, each of which contained a complete femur, tibia (one of the two leg bones found in land animals) and a few foot bones.[60] This should be somewhat surprising, given that the ancestors of modern whales fully lost their hind legs about 35 million years ago. Clearly, only one or at most a small number of mutations suffice to restore hind-limb development in whales; that is, the mutational distance between having and not having hind limbs must be short. The ability to build at least rough-hewn hind legs has been retained in whale GRNs throughout 35 million years of non-use and has merely been switched off. When hippopotamus-like creatures took to the sea, they fully lost their hind legs in perhaps 10 or 15 million years, but two or three times that period has not sufficed for them to lose the dormant GRN relationship structures that sometimes pull partially-formed legs out of an atavistic hat. In the language of attractors and their basins, some GRNs in current whales are camped out near a ridge on the other side of which partial hind legs return.

The case of recurrent frog teeth is even more impressive. About 230 million years ago, the ancestors of modern frogs lost their lower teeth, retaining only a few incomplete teeth on their upper jaws.[61] However, about 10 million years ago, one species of frog, *Gastrotheca guentheri*, regained rudimentary lower-jaw teeth. That's a remarkably long memory for a trait that is itself built of traits built of further sub-traits, such as the association between bone, dentine and enamel. Frogs have "remembered" their lost teeth for fully one-third of the time that animals have existed at all. Although they currently use their partially-formed teeth for grasping rather than chewing, if frogs were to come under selection for an ability to chew, one can easily imagine that their incomplete teeth would climb well-trodden pathways up a fitness peak and rapidly become identical or at least similar to their long-lost true teeth.

But there is a significant similarity, a kind of a cheat, in these two examples: Whales retain forelegs as pectoral fins and frogs still possess a few upper teeth. The patterning modules used to build the remaining features can presumably be relinked in such a way as to rebuild the lost traits. Still, even if this is the source of trait recurrence, it remains notable that an easily restored linkage between front and back limbs or upper and lower teeth persists as a possibility even after one of the linked features has disappeared.

And atavisms can also occur even when patterning modules have entirely fallen out of use. Snakes lost their front legs about 170 million years ago and their hind legs 70 million years ago, and yet they are occasionally born with one or two sets of partial legs. Furthermore, although none survive today, the fossil record shows several examples of snake species that re-evolved apparently functional legs (although they might have used them for grasping rather than walking).[62] This should be somewhat surprising. Adaptations in current use are maintained in good form by selection, but one would expect an unused module/sub-module network to deteriorate through mutational drift. That the relationships between elements of a trait can outlast the trait itself shows the power of developmental/evolutionary memory. Presumably, snake legs first returned in semi-vestigial form but were tuned up to operational status through standard hill-climbing processes.

A close reader may object that atavisms are less like partially forgotten memories and more like memories that haven't been accessed recently. Suppose it has been two decades since I last had to repair a bicycle tire, but suddenly I do; I have not actually lost the memory of how to repair a tire but merely failed to retrieve it for a while. Similarly, one might say that snake GRNs haven't forgotten how to build legs but merely *don't* build them; leg-building GRNs have been switched off. But memory operates on a continuum. When I attempted to repair a bicycle tire last summer, after an interval of three decades, I left out one step (roughening the rubber); similarly, most atavisms show only partial restoration of traits. Memories gradually fade, although some, like the physical knowledge necessary to ride a bicycle (the cerebellar model of bike riding), fade so little that they are commonly believed to remain fully intact.

Evolutionary memory persists over millions or hundreds of millions of years in large part because it is instantiated through redundant relationships. Traits that have been selected together under many different conditions have necessarily overcome many kinds of interfering factors, and to resist that interference, they evolved increasingly redundant Hebbian connections. Friendships that survive war and peace and prosperity and poverty are similarly difficult to break. Even when a GRN crosses over a ridge and *all* the connections between traits vanish, those relationships still lie just on the other side. Redundantly established modules and linkages are attractors with very large basins.

Related mechanisms preserve memories of the ordinary kind.

The Persistence of Memory

It is a matter of common experience that once associations are formed, they are not easily lost. Dogs that learn to salivate at the sound of bells or to get excited at the donning of a coat can unlearn the habit, but the process of associative "extinction" is slow. In part, memories are held in place through redundant pathways. To see how this works, consider PTSD in military veterans.

PTSD symptoms include intrusive memories that include intense anxiety and panic. These memories are typically formed through repeated traumatic events. For example, veterans of the 2003-2011 Iraq War were heavily traumatized by frequent encounters with the disguised roadside bombs referred to as improvised explosive devices (IEDs). Any sensory input that frequently occurred in the temporal vicinity of an exploding IED became linked to the experience in the typical Hebbian fashion. These included unexpected explosive sounds and the sight of ordinarily innocent objects, such as soda cans lying on the ground.

For some veterans, sensory inputs reminiscent of those associated with IED explosions retain an ability to trigger full-blown traumatic memories even years or decades later. This deeply problematic and sometimes disabling persistence of memory constitutes PTSD.

To reduce their PTSD symptoms, patients must find ways to weaken or erase the Hebbian connections that produce them. One set of therapeutic techniques uses repeated deliberate exposures to memory triggers (or at least verbal invocation of them) under conditions in which the veteran otherwise feels especially safe and comfortable. Unfortunately, the separate "vertical" Hebbian associations that produce PTSD symptoms are also connected "horizontally," and when memories are triggered along one pathway, they may strengthen or restore others. Suppose that the Hebbian connection that associates innocent-looking objects lying on the side of the road with ferocious explosions has begun to weaken, but then a different trigger, such as a backfiring engine, initiates a PTSD panic response. Once in a state of panic, the PTSD sufferer may call to mind multiple other occasions of panic, some of which were, in fact, preceded by the sight of a roadside IED. The moment those visual images are recalled, their Hebbian linkage to terror and panic will be partially restored.

Even people without PTSD experience intrusive recollections that they would prefer to forget. Webs of memory can be terribly sticky.

But there is a positive side to the associative nature of memory too. I have a friend who lived in Finland until she was seven and then moved to the US and forgot Finnish almost entirely. She remembered at most a dozen words. Later in life, she decided she wanted to recover her ancestral tongue and enrolled in a Finnish language class. She rapidly surpassed all the other students, not so much because she remembered specific words, but because she seemed to intuitively understand the structure of Finnish; the way the words fit together in sentences simply seemed natural to her, although it didn't to the other students. The source of this intuition and sense of naturalness was her unconscious recall of word *relationships*.

Systems that operate in the fashion of neural networks hold their information at multiple levels and in a distributed fashion. Metaphorically speaking, such memory is holographic, and almost any portion of the whole represents the whole.

Analogous processes sustain Hebbian linkages in GRNs and permit them to be recovered and restored even when they have been partially lost. When a set of linked adaptations arise and are sustained under numerous environmental conditions, the linkages that forge them become deeply intertwined and difficult to break, and if one Hebbian connection is restored under selection, it may activate some of the others through horizontal linkage.

Some redundant connections act as catchalls or backstops, sweeping up miscellaneous details even when the specific processes originally designed to manage them fail in whole or in part. These backup methods may not be as accurate as the primary ones, but they do the job, or at least they do it well enough for more accurate systems to re-evolve. To see how this can work, let us return to neural networks and consider how ground-dwelling wasps find their way home.

Ordinarily, wasps find the entrances to their nests by recalling the general direction to them and then searching for familiar landscape features along their return flights.* However, when obnoxious scientists erase local features, wasps fly further upward and identify the location by triangulating on somewhat more distant objects.[63] If intermediate-distance objects are also disturbed, these much-put-upon wasps simply move even higher up. Their ability to memorize geographic details at multiple levels is rather remarkable for a creature with relatively few neurons.

More fundamentally, the inherent structure of biological modularity can help restore damaged evolutionary memories. All the various general-purpose widgets and sub-widgets utilized by an organism represent *methods* of decomposing the general problem of survival into discrete adaptations, and those forms of decomposition themselves contain information. The various pieces of an animal fit together, and the presence of a few implies all the rest. Even if one or more elements of a set of linked adaptations are lost, the problem's basic decomposition remains intact and ordinary hill-climbing can often reproduce the missing pieces.

For an illustration of this idea, imagine that a group of Neolithic humans stumbled across an abandoned cart with one of its two axles broken. Even if this group was somewhat technologically backward and had no prior knowledge of wheeled carts, it seems plausible that

---

* Many insects, including ants, bees and wasps, can set off from point A following a zig-zagging track, and then sum up all the zigs and zags to calculate a direct route back, an ability called "path integration."

they would soon recognize how the cart was meant to operate and re-engineer an operational one. Evolution is far less intelligent than humans, but even so, fewer mutations will suffice to restore the missing elements of a damaged set of adaptations than to engineer an entirely novel solution to the problem those adaptations once addressed.

The Hypothetical Evolution of Land Dolphins

Let us put all the above memory factors together and work through how dolphins might evolve if they came under selection to return to life on land.

Dolphins resemble whales in that they are sometimes born with hind-leg atavisms and possess a residue of past legs in the form of pelvic bones that have lost their connection to the spine. From these facts, one may plausibly conclude that they retain elements of the GRNs that once built fully functional legs. Perhaps only a few switches have been turned off, and remnants of the original patterning modules still exist; if so, those pelvic bones could serve as a template for gradually reconstructing a full limb through a limited number of hill-climbing steps. Or perhaps the widget that currently produces pectoral fins could reconnect to the damaged widget that now produces only vestigial pelvic bones and, by backstopping its function, convert those bones into a second pair of fins; as a next step, various modules that produce limbs rather than fins could revive in both, augmented by spontaneous hill-climbing. No matter what the route, dolphins under selection for land living could almost certainly regain both front and back limbs far more quickly than if they had to reinvent them from scratch.

Dolphins going back to the land might also rapidly re-evolve hooves. Animal surfaces that typically rub or strike against the external environment are typically reinforced or protected by structures made of the protein keratin; as mentioned above, examples include skin, fur, hair, hooves, claws, nails, horns, antlers and teeth. The relationship is instantiated in a kind of deep connection between the processes that construct the external surfaces of an organism and the processes that manufacture and weave keratin. Both keratin itself and the associations between keratin and external contact are thus remembered by evolution, and this connection might help them rapidly re-evolve keratin at the ends of their newly developing legs, too.

Caveat: I do not mean to suggest that evolution re-creates the exact macroscopic anatomic features of ancestral organisms. The legs of land dolphins would resemble those of their land-dwelling ancestors but would likely evolve at least somewhat differently. Furthermore, there is no reasonable chance that the *entire* animal once more wading in ponds and heaving itself over soil would exactly duplicate its land-dwelling ancestors. Evolution never exactly repeats itself in specifics; what it repeats are *patterns*. An animal is a collection of features that address environmental conditions, and when environments change along a particular dimension, adaptations can typically respond along that dimension alone while leaving (most) other features unchanged. The result is a novel combination of traits.

To give a simplistic example, giant anteaters, aardvarks and elephant shrews are insectivores with long snouts that evolved from ancestors with shorter snouts. Let us plausibly suppose that all three of these evolved their long snouts to expedite feeding on various insects; less plausibly, let us imagine that there is a single switch that, when activated, causes an animal's snout to elongate and narrow. (Snout length is more plausibly controlled by a continuous dial; nonetheless, as will be discussed shortly, the very existence of dials may reflect past evolutionary experience of frequent switches along the dimension controlled by that dial.) Despite the use of the same "remembered" capacity, the total collections of traits that constitute a giant anteater, aardvark or elephant shrew are quite distinct even though they lie close together along the dimension of snout shape.

Just as animals repeat patterns, so do Earth environments; what has occurred before often occurs again, although with variation in detail. When environments frequently switch between structurally similar states, organisms may develop efficient ways to switch between responsive adaptations. To see how this works, let us look at a form of recurrent adaptation more subtle than the lengthening of snouts or the reconstruction of legs.

Developmental Memory and Expedited Mutational Pathways

GRN Hebbian linkages encode relationships between traits, sub-traits and sub-sub-traits. These relationships presumably include various size, length, weight and distance ratios. We have already encountered the 1:1 match of bilateral fin lengths, but there must be many others that are somewhat more subtle, such as (perhaps) optimal ratios that relate leg length, body weight and the distance between front and back legs.

Let us suppose that optimum limb-length and leg-distance ratios are different in light, fleet-footed deer-family creatures such as white-tailed deer than in heavier, lumbering ones such as moose or elk. Let us also suppose that a single developmental "dial" controls total body size, but that limb length and distance relationships must be adjusted individually and less easily. If, over the course of evolutionary time, the descendants of a deer species periodically oscillate between the overall size of a moose and of a white-tailed deer, during each such "easy" transition in total body *size*, these bodily *proportions* will have to undergo a one-by-one optimization process that in the end achieves the same ideal values. After sufficiently many of those transitions, Hebbian relationships setting the values for those proportions will develop a persistent footprint in GRN modularity. From then on, whenever a moose-sized deer comes under selection for reduced size, the appropriate proportions will appear as if by magic. But there would be nothing magical about it; the Hebbian relationships in a deer's GRNs act as a model for various possible body sizes and associated proportions, and a relatively small number of mutations can suffice to restore the proportions that go with total size. This (hypothetical) rapid reconstitution of frequently useful ratios illustrates a form of developmental memory that is a bit more subtle than the simple reactivated switches that produce atavisms.

Another memory phenomenon can be observed in the emergence of expedited mutational pathways along dimensions of common, recurrent change. When environmental conditions frequently shift back and forth between two or more states, recurrent re-evolution of the same adaptations may induce the formation of developmental processes that bring these adaptations near to one another in mutation space (Parter et al. 2008). This phenomenon has been observed in simulations and is also easy to explain intuitively.

Each time environmental conditions recur, some members of the population will possess GRNs that happen to be structured in such a way that the necessary adaptations can be achieved with relatively few mutations. The descendants of those organisms will reach the adaptive goal first and will therefore have a chance to flourish and dominate before the others arrive. If conditions shift back and forth repeatedly, and at a rate rapid enough so that no one condition becomes locked in place, selection will tend to find organisms whose GRNs place adaptations to those recurrent conditions close together in mutation space. To quote Tobias Uller and his coauthors,

> [Computer modeling indicates that] switching between two environments at a frequency that enables populations to adapt but not to evolve regulatory networks that are mutationally robust, will tend to push genotypes toward a space of possible regulatory networks where the mutational distance is short between networks that are functional in environment one and networks that are functional in environment two (Uller et al. 2018, p. 956).

This reduced mutational distance is itself a form of memory, but of common evolutionary *trajectories* rather than structures.

Reduced distance can emerge in the form of switches that cause a sudden jump from one phenotypic form to another, such as the automatic reemergence of a complete suite of modified limb proportions or (less plausibly) an elongated, narrow snout. However, it can also manifest as a parameterized system, a dial or set of dials that makes continuous shifts along one or more dimensions mutationally easy.

Consider the evolution of the systems that many birds use to build their varied beaks. The history of bird evolution includes frequent changes in food sources that must be addressed through alterations to beak shape. Depending on what birds eat, they may need beaks that are long or short, narrow or broad, or deep or shallow. Perhaps each beak type was once constructed through its own idiosyncratic process, but today, many and perhaps most bird species build their various shaped beaks by adjusting just a few continuous parameters.[64] This makes it easy for them to evolve appropriate-shaped beaks when available foods change. Similar processes permit easy evolutionary variation in tooth shape (Uller et al. 2018; 950-951).

Parameterization of this kind can be regarded as a form of self-modeling because it records the fact that beaks have often varied in a continuous fashion along one or more dimensions; repeated evolutionary adaptation back and forth along those pathways has converted them from faint trails to Roman roads.*

We have now justified (in an intuitive sense) Watson's second claim, that Hebbian connections and other forms of self-modeling strengthen and maintain "memories" of past adaptations. Watson's theory goes further than this, but let us take a moment to explore how "mere" developmental/evolutionary memory can enhance future evolvability.

---

* Trait parameterization is not usually included as an element of facilitated variation, but perhaps it should be. It certainly makes adaptation more efficient along the parameterized dimensions.

Developmental/Evolutionary Memory Facilitates Variation

If an adaptation was found useful in the past but was subsequently abandoned, it stands a good chance of proving useful again if the conditions under which it first evolved return. We have already discussed this in the hypothetical evolution of land dolphins. For a more realistic example, and one that may be relevant today, let us briefly consider the history and future of corals. We will return to corals in more detail near the end of this essay.

Corals are marine invertebrates that build colonies of genetically identical clones. As they grow, they secrete calcium carbonate—basically limestone—and use it to build hard exoskeletons. Thousands or millions of these colonies create reefs such as the Great Barrier Reef of Australia.

Corals have persisted in much the same lifestyle for hundreds of millions of years. From time to time, the ancestors of modern corals have lived in environments as hot or hotter than those that global climate change is expected to produce; during other eras, they endured temperatures colder than they are today. During each environmental condition, evolutionary processes operating on Timescale Two optimized adaptations to that condition. Because corals have sequentially adapted to many environmental conditions, they have used and perfected many effective adaptations over Timescale Three.

The adaptations employed by corals living today are appropriate for current conditions, but the GRNs that control their developmental processes presumably retain memories of past adaptations. This evolutionary memory may help them adapt to anthropogenic climate change. It is possible and perhaps likely that corals can take a past set of adaptations for warm seas off the shelf and rapidly put them to use. If so, they may adapt and bounce back more rapidly than expected as global temperatures rise.

Optimization on Timescale Three Guides Optimization on Timescale Two by Modeling the World

> *"Evolvability is to evolution as generalization is to learning."*
>
> WATSON AND SZATHMÁRY 2016.

We are now ready to look at Watson's third and most daring proposal: that the Hebbian structure of GRNs permits them to facilitate useful variation by generalizing from past experience. While the evidence for this theory remains incomplete, what Watson has been able to demonstrate so far hints at a revolutionary new paradigm for understanding evolutionary processes. I will begin with an overview based on a general analogy to neural network learning systems, and then present simulations and intuitive arguments that explore the proposed processes in more granular detail.

Simple Hebbian relationships represent simple generalizations about relationships between organisms and the world, such as that bilateral fins should match to make swimming efficient. More complex Hebbian relationships represent more subtle generalizations, such as the optimal ratios between body weight, leg length and distance from front to back legs in deer-family creatures of a certain weight. Because many low-level adaptations used today first emerged in prokaryotes, and another large set in unicellular eukaryotes, the hierarchical depth of the nested associations used by developmental processes is immense. Low-level associations represent fine-grained details of the world and higher-level ones track how those details relate to one another. The deeper the hierarchical stacking, the greater the potential subtlety of the model built out of them. Thus, the Hebbian interrelationships between adaptations embodied in GRNs can, in principle, model the world at a profound level of detail.

The power of complex Hebbian relationships underlies the success of large language models, of which ChatGPT is the best known at the time of this writing. While ChatGPT uses a "transformer" neural network architecture and GRNs are constructed as hierarchical Hebbian networks, the two architectures share deep similarities, and each may be able to provide insights into the other.

Through a process of training on massive quantities of text, large language models learn the deep patterns that underlie the language in which that text is written. This knowledge not only permits them to construct meaningful and responsive sentences, paragraphs and texts but also to recognize the subtle characteristics of specific styles and genres. A given set of words may be a poem, an advertisement, a polemic, or a work written by Shakespeare or Hemingway, and a large language model can tell the difference.

This ability shows itself in startling fashion when such a model is asked to ventriloquize an authorial voice, for example, to predict how *Hamlet* would sound had it been written by Hemingway. Words vary widely between texts, but the relationships between the words found in texts and the relationships of those relationships remain somewhat constant, or at least constantly characteristic, within subjects, genres and the oeuvres of authors. When a large language model composes its language, it does not quote the text it has trained on in the manner of evolutionary memory. Instead, it generalizes, using its deep knowledge of textual relationships to recombine words in novel ways that nonetheless match the required style.

Here is how ChatGPT4 rewrote the above paragraph when prompted to adopt the style "standard popular science:"

> Step into the world of artificial intelligence and you'll find a tool not unlike a chameleon of the written word—a large language model. This AI isn't just any tool; it's a master of mimicry, capable of transforming texts with the finesse of a seasoned author. Challenge it to reimagine the introspective soliloquies of *Hamlet* as if penned by the sharp, concise Hemingway, and watch the magic unfold. This transformation isn't

about swapping words haphazardly. Like a scientist finding patterns in chaos, the AI detects an underlying structure—a DNA of literature, if you will. The essence of how words connect, the syntax, and the stylistic flair are preserved, even when the words themselves are in flux. It's this consistent literary fingerprint that the AI preserves, adhering to the invisible rules unique to genres, topics, or the signature of a specific author.

And here is ChatGPT4's rendition of how Hemingway might say it:

If you ask a big language machine to take some text and change how it sounds—to make Hamlet talk the way I would—it can do it well. Words change from one piece of writing to another, but the way they fit together doesn't change much.

The popular science version is pretty good, if flowery, but ChatGPT's Hemingway missed the crucial point that genres and authors utilize characteristic patterns all their own.

Watson proposes that the deep Hebbian structure of GRNs gives them capacities somewhat analogous to those of large language models; however, instead of rewriting texts to match different styles, they partially rewrite organisms to match new environments. When "prompted" by novel Earth conditions, GRNs are so constructed that random mutations have an elevated probability of producing responsive adaptations, including novel *sets* of adaptations that have never appeared in that organism's evolutionary history. (This is a more subtle version of the tale of the multiple distinct organisms that all elongated their snouts.)

To understand how this could work, let's start with evolutionary memory and build up from there.

At a minimum, established Hebbian relationships should cause mutations to preferentially test traits in combinations or relationships that have worked well in the past rather than trialing combinations or relationships that have frequently failed. For example, evolution has learned that claws belong at the end of limbs rather than emerging like epaulets on the shoulders.

More subtly, if certain environmental conditions typically shift together, the modules that produce traits to address those conditions tend to become linked. Suppose (hypothetically) that for a certain plant species, higher spring temperatures are consistently associated with increased predation by leaf-eating insects in summer. Such a plant will evolve separate adaptations to high temperature and leaf predation but may also profitably build in a linkage between them; by doing so, whenever temperatures rise in spring, the plant will not only deal with the temperature rise itself but also prepare defenses in advance against the expected forthcoming insect attack.

Suppose next that climate conditions shift toward lower temperatures for many million years and a plant species loses those adaptations. If average temperatures subsequently rise again, the persistence of Hebbian memory will increase the probability that both adaptations re-emerge together; a mutation that activates one will activate the other, too. This mechanism serves as a form of enhanced evolvability.

More complex sets of correlated and independently varying Earth conditions are similarly reflected in correlated and independent GRN relationships. As a result, the pathway for easiest phenotypic change through mutations tends to match "the natural decomposable structure of the [past] selective environment" (Kouvaris et al. 2017).

Here, we are approaching something like what ChatGPT achieves when it learns the structure of texts. Instead of identifying which words are likely to follow other words, Hebbian structures in GRNs identify which adaptations are likely to go well (or not go well) with other adaptations. Even more subtly, they encode the many different ways that adaptations have related to one another in the past when they arose in response to various environmental conditions. These overall patterns can be understood as the evolutionary equivalent of genre or authorial style.

Of course, rather than words and their relationships, Hebbian networks in GRNs represent skeleton-key widgets and the relationships between them. GRNs thus encode the correlational structure of the ever-changing world in the form of adaptations that themselves encode generally applicable facts about the world. Taken as a whole, such a system "knows" a tremendous amount about what has worked well in the past. And knowledge of the past can suggest appropriate responses in the present.

Conditions on Earth do not tumble about chaotically but exhibit regularities on many levels. Day follows night, winter follows summer, and it is almost always warmer in the sun than in the shade. If there were no regularities, brains would be useless and would never have evolved, but nature is drenched with order at all scales of size, time and energy level. The Hebbian-structured neural networks that constitute brains model those regularities and, when faced with novel conditions, generalize from past experience to propose appropriate responses. Similarly, GRNs encode many regularities of the world in the Hebbian correlational structure that links their adaptations; when a population of organisms faces a novel environmental condition with structural similarities to past conditions, that Hebbian structure, in effect, generalizes from the past and offers up variations that have an elevated likelihood of increasing fitness in current conditions.

We have now reached the quote beneath the title of this section: "Evolvability is to evolution as generalization is to learning" (Watson and Szathmáry, 2016). ChatGPT has learned to generalize about language and uses that ability to process and produce novel texts when prompted to do so; humans generalize from past experience when considering how to address current issues; general-purpose widgets embody generalizations in their structures and processes; evolving GRNs have learned how to combine and recombine general-purpose widgets to efficiently innovate responsive variations when environments change in "structurally recognizable" ways. The ability to use past evolutionary experience to facilitate current adaptive variation makes organisms more evolvable.

Or, at least, that is the prediction of Watson and his associates at the University of Southampton. Other contributions in a similar vein have been made by N Kashtan, S Ciliberti, M Parter, U Alon, G Wagner and others.[65] These researchers in the related fields of computational biology and artificial life perform computer simulations of evolutionary processes and have concluded that the self-modeling characteristics of GRN Hebbian associations permit them to guide future variation through inductive generalization on the past. Although this research uses simplified models that cannot yet capture the full range of evolutionary processes, the results are robust and impressive. (For a good technical overview with numerous supporting references, see Watson et al. 2016.)

This is the full theory of natural induction through self-modeling that I have been teasing throughout the above.

The remainder of this section approaches the subject of evolutionary generalization through Hebbian relationships from several directions and in several ways. Watson and his colleagues have devised two clever simulations for the express purpose of making their difficult ideas easier to understand, and I describe these in detail. In addition, I offer two purely intuitive explanations that express various elements of the theory. Hopefully, by the time you reach the end of this essay, you will agree that their extraordinary proposal is, at the very least, conceptually persuasive and worthy of further study.

Brief Intuitive Explanation

The generalizing power of Hebbian relationships is obvious when it operates in animal or human brains. To give a simplistic example, suppose deer lived on a continent without wolves, and then wolves were introduced. Because they had never seen wolves before, these deer might take some time to inspect them but would soon notice their sharp teeth, assess their size as large enough to be dangerous, and perhaps wait around long enough to observe the wolves beginning to run toward them. But it wouldn't take long for them to classify wolves as potentially dangerous predators. Once having made the classification, they would activate a highly general-purpose behavioral widget used by deer and pretty much every other mobile organism: running way.

The rule "Run for your life!" is a generalization over threats, and it's a good one. But that top-level rule contains further sub-rules, such as, perhaps, "If the predator is gaining on you, try zig-zagging or aiming for difficult terrain." At a deeper level, deer possess an advanced internalized model that allows them to rapidly traverse ground strewn with rocks and roots while at the same time dodging or overleaping protruding branches and to solve all the necessary differential equations at speed. Running through a forest requires tight integration of sensory input and physical action with little time for calculation, and it only seems easy when deer do it because they are incredibly skillful. The higher-level behavioral model "escape" contains multiple sub-behaviors that can be mixed and matched, tweaked and recombined; it is an effective model of effective models.

When a deer escapes a predator by running through a forest, it can't exactly reuse any prior strategy in the manner of developmental memory; forests vary, as do the capacities and strategies of individual predators and the relative initial positions of predator and prey. But deer brains do not literally memorize past escape trajectories or limit their vigilance to known threats; instead, they model a system for escaping that is applicable to most environments. This modeling, too, is a form of generalization.

The deer model of "how to run on uneven terrain" was built through Timescale Two evolutionary processes supplemented by within-life practice operating on Timescale One. Deer are born knowing that they should flee danger and more or less how to do it; nonetheless, young deer must practice their running skills. Fawns get the zoomies just like puppies and learn the tricks and pitfalls of running by tripping, falling and crashing. However, their learning processes are guided by the invisible hand of inborn capacities. Fawns don't need to discover how to run by first trying wacky motions à la Monty Python's "Ministry of Silly Walks," nor need they practice in all or even very many possible configurations of terrain; they are hardwired with a model that represents successful flight, and when foolish fawns zoom about, the things they think of trying already fall along the lines of what they will need to perfect.

According to Watson's theory, the Timescale Three models instantiated in Hebbian structures similarly guide Timescale Two processes by tending to place variations likely to succeed nearby in mutation space and placing adaptations likely to fail further away. On the simplest level, Hebbian associations bias variations toward plausible options (ones that often have worked in the past) and away from nutty options (ones that have seldom worked well in the past, like claws attached to shoulders instead of paws). More subtly, deep hierarchical relationships reduce the mutational distance to *patterns* of adaptations that have proven successful for conditions that are in some sense structurally similar to those that are present today.

Thus, the Hebbian relationships encoded into developmental systems, like the inbuilt model of "how to run away" possessed by deer, enable evolution to limit grinding trial and error and reach novel optima with relative ease.

Let's look at some of the simulations offered by Watson and his colleagues in support of these claims.

Love the One You're With

The following discussion presents, albeit with some poetic license, a simulation published by Watson and his associates under the delightful title "If you can't be with the one you love, love the one you're with" (Davies 2011). They offer this amusing simulation for the express purpose of making difficult ideas easier to understand; it is a simplified version of the more sophisticated computer simulations that underly Watson's theory. I will first present this intentionally simplified simulation and then briefly describe the more realistic ones.

Like the computer simulations mentioned earlier that Watson and his colleagues used to demonstrate the power of developmental/evolutionary memory, this one utilizes a set of pairwise Hebbian relationships with no hierarchical depth. In effect, it models the evolution of a simplified prokaryote in which many genes can be divided into two groups (operons) that cause them to be turned on and off together. In this simulation, the action of a form of Hebbian association permits a group of interacting virtual agents to efficiently find near-optimal arrangements. It seems plausible that more complex Hebbian relationships could accomplish much the same thing in more complex variational spaces.

As noted in the paper, it is easier to describe the simulation than to explain why it works; facilitated optimization seems to leap out as if by magic.

Let us suppose that there are 100 brilliant but sensitive and temperamental computational biologists working in the Computational Evolutionary Biology department at the University of Southampton. Some of these scientists work well together in pairs; they find one another's critiques helpful, and when they start speculating together, they frequently come up with good ideas. They increase one another's productivity. However, not all scientists have good chemistry. Some pairings are like bad marriages; one tends to stunt the other's ideas by cleverly and insidiously, or sometimes directly and rudely, dismissing their most promising lines of thought.

These working relationships are reciprocal; if A gets along with B, B also gets along with A. However, they are not transitive; A may get along well with B, and B may get along well with C, and yet A and C may go at each other's throats in seconds.

Each scientist knows which of the other scientists they prefer to hang out with. However, there are only two possible places where they can meet to discuss things as a group: Auditorium 1 and Auditorium 2. Within each auditorium, scientists will take turns standing up and addressing the group, providing helpful responses to the ideas of their well-matched counterparts and toxic, undermining responses to those with whom they are incompatible.

At the beginning of each workday, the scientists are randomly assigned an order and allowed to choose their auditoriums in sequence. The complexity of pairwise interactions causes this choice process to demand one compromise after another.

Each scientist assigns helpful relationships a value of +1 and harmful ones a value of –1. Scientists assess the compatibility value of each auditorium by adding up the good and bad relationships in that auditorium. They choose an auditorium by picking the option with the highest summed value.

Let us imagine that the sequence chosen for the first day is alphabetical. Scientist A picks an auditorium at random. Since Scientist B gets along with Scientist A, B chooses the same auditorium. Since C works well with B but not with A, either auditorium offers C a compatibility value of 0, either straightforwardly if C chooses the empty one, or by summing the values for A and B if C elects the populated auditorium. Let's say C chooses the empty auditorium.

At this point, A and B are in Auditorium 1, and C is alone in Auditorium 2. Let us now suppose that scientist D gets along well with A but badly with B and C. D's best choice is Auditorium 1 because its compatibility value for him is zero, while the compatibility value of Auditorium 2 is negative one.

This process continues as each scientist in succession chooses the best or, perhaps, the least bad auditorium. Once all scientists have made their initial choices, they then make a second round of choices in the same order, this time deciding whether to stay where they are or to switch. They repeat this sequential decision process until they reach equilibrium, a distribution of scientists that cannot be improved by further switching. (It can be demonstrated that this will happen eventually.) The final arrangement is a local optimum, the most productive arrangement that can be achieved through hill-climbing from the initial conditions (the random order assigned to the scientists).

The overall quality of the arrangement achieved at the end of the switching process is then measured by summing all the compatibility values experienced by each scientist, yielding a final "compatibility sum," reflecting the total productivity of the arrangement. If each scientist were able to end up in an auditorium populated only by compatible scientists, the total compatibility sum of this blissful arrangement would be $100 \times 100 = 10{,}000$. This figure could be achieved if, for example, all members of the first 50 got along with one another and chose Auditorium One, and all members of the second 50 were compatible and chose Auditorium Two.

Alas, the overwhelming majority of possible bilateral relationships chosen at random include complex interfering relationships that preclude such happy outcomes. For any given set of scientists and their relationships, there is an optimal arrangement for maximum productivity, the equivalent of a global fitness peak, but the process of sequential room choice will (almost) never find it; instead, it will (almost) always get stuck on a local optimum, a hillock or a knob far from the peak.

Before beginning their simulation of room-switching scientists, Watson and his coauthors used a randomization process to set pairwise relationships between all the scientists. Once set, those values were never changed. However, before each simulation began, the *order* in which the scientists would make their sequential choices was randomly shuffled.

If the best possible compatibility sum for a set of scientists with a particular set of compatibility values is (let us say) 3,000, this number could be achieved by telling each scientist which room they should choose in the initial round. However, the actual process didn't permit such omniscient decision-making; instead, the scientists were compelled to go in sequence and at each opportunity given them to choose an auditorium to seek only to maximize the compatibility value for themselves.

Through 1,000 simulations, the achieved compatibility sum never approached 3,000 but hovered around a much lower local optimum of, let's say, 1,000.

Now comes the interesting part. An additional 1,000 trials were run using a slightly altered methodology, one that added a Hebbian-like "familiarity factor" into the mix. At the end of each completed room-switching process, scientists who ended up in the same room were made to become somewhat more tolerant of one another, even if they remained fundamentally incompatible. This effect was achieved by adding a small familiarity increment of (let us say) .001 for each pair each time the final equilibrium state of a room-switching round placed them together. The value of the increment was set sufficiently small to ensure that the familiarity adjustment occurred more slowly than the room-switching process. The familiarity increments determined at the end of each round were added to a familiarity-factor matrix to keep a running total.

In this second set of simulations, each time scientists choose which room to go to, they consider the fixed compatibility factors PLUS the small but steadily growing familiarity factors. If they often find themselves sharing a room with the same scientist at the end of switching rounds, the total familiarity factor between them may grow large enough to counter poor compatibility values; pairs that already arise frequently will arise even more frequently and those that already arise rarely will tend to arise even less often.* The system of scientists is, in effect, learning facts about its own behavior.

For example, suppose that at one point in a series of room-switching trials, A has ended up in the same room as C 20 times. In the next round, A and C will each soften their -1 antipathy by $20 \times .001 = .02$, producing an effective antipathy of $-.98$. If A ends up with C 800 times,

---

* This can be thought of as a form of action against selection, reminiscent of the prior discussion of how the dead hand of the past may assist the discovery of general-purpose widgets, and how a form of evolutionary friction (mentioned in a prior footnote) may favor certain kinds of phenotypic plasticity.

this adjustment factor will reach 0.8, almost enough to fully overcome their innate incompatibility.

Importantly, these familiarity factors do not alter how the total compatibility sums at the end of a completed room-switching cycle are computed; they only play a role each time a scientist *decides* which room to pick in the first place or to switch to afterward.

The researchers ran this modified system 1,000 times, allowing familiarity factors to accumulate, and found that as the simulations repeated, the scientists began to come to rest at increasingly good compatibility sums, coming closer to the best possible score of 3,000. In other words, the ability to remember and favor the pairings that had frequently occurred in past local optima caused the system to achieve progressively better local optima.

The researchers then fixed the familiarity factors in place and ran the simulation another 1,000 times. The average achieved compatibility sums of these 1,000 trials clustered at a much higher level than the poor achieved compatibility sums found through the first 1,000 trials without habituation. Memory of past correlations in the form of Hebbian associations enhanced the effectiveness of simple adaptive hill-climbing and allowed the entire set of scientists to become as productive as possible.

It is important to note that the final optima achieved by this process consisted of combinations of scientists that were never achieved prior to the addition of the familiarity effect. In some way, the ability to recall characteristics of past local optima permitted the discovery of new, much better local optima.

It is not easy to explain why including familiarity factors made such a big difference, and the following intuitive explanation only captures part of it.

If A and C end up in the same auditorium at the end of a high percentage of switching cycles, it means that auditorium cohabitation by A and C is a feature of many local optima. Because of this, there is an elevated chance that A and C will share an auditorium in more nearly global optima, too. Another possibility is that A and C might typically briefly share an auditorium on the way toward an excellent outcome, but they often end up stuck together at the end of room-shifting cycles because other poorly arranged scientist pairs block further optimization steps—a blockage that familiarity factors remove. In either case, familiarity factors model neighborhood characteristics of local optima, and that model possesses information about optima in general. Thus, local information derived from a set of past optimizations guides the way toward achieving novel optima that were never sampled before. To put it another way, the system of room-switching scientists, given the additional capacity of familiarization, generalizes from past experiences to find novel, improved solutions.

Even without the use of familiarity factors, room-switching scientists would sometimes reach excellent configurations, but only rarely. They could do far better if they were permitted to consult with one another and agree to make several uncomfortable temporary switches on the way to better configurations. However, to reflect how actual evolution works, the experimental setup limits scientists to room switches that provide immediate benefits. Because of this, they often get stuck on poor optima, the equivalent of poor fitness peaks. The addition of familiarity factors converts U-shaped paths that dip before rising into ones that slope steadily upward and thereby provide a path to better optima.

The analogy to evolution is straightforward.

When a population of organisms is evolving in the face of an environmental challenge, mutations "experiment" with the effect of changing various traits. However, traits interfere with one another in complicated ways. Some work well together while others do not. When a population "tries" to optimize traits, it must struggle with incompatibilities and tradeoffs and attempt to achieve the best combination. Evolutionary adaptation is further hampered by the fact that all variations that come under selection must lead to immediate improvements, or at least not cause harm, because evolution can't indulge in four harmful changes in hopes of finding a fifth that will redeem them; it can't easily climb down from local fitness peaks.

The multiple cycles of room-switching reflect the multiple occasions during which evolutionary adaptation, operating on Timescale Two, does the best it can with what it's got. The addition of a slowly increasing familiarization effect models the Hebbian relationships that form and persist over Timescale Three, causing traits that frequently end up together to become more likely to emerge together. The results of the simulation suggest that the addition of a certain favoritism for gene relationships that have occurred frequently in the past may allow evolutionary adaptation to discover new and better gene combinations—even ones that have never been sampled before.

The model used in this pedagogical simulation is highly simplified, limited to −1 and −1 compatibility factors rather than the sliding-scale range of interactions that genes actually engage in. The limitation to two auditoriums is also artificial; even bacteria use hundreds of gene groupings that switch on and off together. However, in other simulations, Watson and his group studied more subtle gene interactions in populations of virtual organisms evolving through shifting virtual environments and found the same results: Self-modeling Hebbian associations permit evolving organisms to in effect generalize from experience and efficiently discover novel, adaptive trait combinations that had never been tried before. (See, especially, Kouvaris et al. 2017 and Watson et al. 2016.)

However, even the most sophisticated of these simulations were "flat," limited to pairwise, horizontal Hebbian relationships between discrete, idealized "traits."* While this may somewhat realistically represent the regulatory systems used by prokaryotes, what we are trying to explain here is the rapid evolution of eukaryotic and especially multicellular organisms, and these utilize hierarchical modular architectures with both horizontal and vertical linkages. Hierarchical modularity is a fundamental necessity for complex organisms. Without it, there can be

---

* In his PhD thesis, Frederick Nash took a stab at simulating hierarchical Hebian relationships. Full text available at *https://eprints.soton.ac.uk/467522/1/Final_Thesis.pdf*

no embryonic development with its compartments, sub-compartments and shared patterning modules. Thus, simulations based on pairwise Hebbian associations employ a significant simplification of the systems they attempt to model.

I asked Watson why his group didn't attempt to model hierarchical modularity, and he explained that with each additional level of hierarchical depth, the dimensionality of the space to be searched increases exponentially.[66] It might be the case that nothing much could come of such a simulation unless it more nearly approached the scale of search that produced multicellular organisms: trillions of virtual organisms reproducing through tens of billions of generations. Unlike the creators of ChatGPT, artificial life researchers do not have the funds necessary to pay for such massive computational power.

Instead, they have approached the problem from the opposite direction and shown that even simple, non-living systems can learn to generalize through self-modeling. This remarkable finding suggests that such processes may occur throughout nature and should be considered ordinary rather than surprising.

Natural Induction Through Self-Modeling in a Physical System

Self-modeling is a form of induction that, if translated into an explicit proposition, might state, "What I frequently found successful/useful/optimal in the past is likely to prove successful/useful/optimal in similar situations today." By recalling their past behavior and generalizing from it, GRNs and brains streamline (optimize) their current optimization processes.

Remarkably, self-modeling may optimize optimization processes in non-living systems, too. In a paper available only as a preprint at the time of this writing, Watson and his group modeled a purely mechanical system whose self-modeling behavior enhances its ability to self-optimize. To traditional biologists who wish to ground everything in natural selection, Watson offers the results of this experiment as proof of principle for the presence of additional ordering forces in evolution—and in the universe more generally. He invented the term "natural induction" to describe these results, which are striking and suggest a new and potentially revolutionary paradigm for studying complex systems.

The system they mathematically simulated consists of 300 virtual anchors each of which is connected to 50 percent of the others by identical virtual springs.[67] To prevent endless oscillations, each spring is damped (constructed in such a way that all oscillatory motions die down in a relatively short time). In addition, these somewhat ghostly virtual springs and anchors are permitted to pass through one another.

When such a system of anchors and springs is placed in an arbitrary starting position, the system "attempts" to reduce strain as much as possible by shifting the positions of the anchors. However, since one spring's relaxation is another spring's stress or strain, the best achievable result will be a compromise. The effectiveness of that compromise can be measured by summing the squares of the distances each spring is stretched or compressed beyond its resting point; the lower the total, the more optimum the result. This "stretch and strain sum" plays an identical role to the final "compatibility sum" in the room-switching scientist simulation. Similarly, the physical tendency of a spring to relax mirrors the efforts of scientists to spend time in a room with most friends and the fewest enemies. Or, to compare this dynamic system to the action of natural selection, differential potential energy here plays the role of differential fitness.

In the scientist simulation, the best possible optimization achievable from a given set of initial conditions was unlikely to approach the true global optimum of the system, and the same is true here. If an outside agent were to calculate the true global optimum and set the system of springs and anchors in an initial state that closely resembled it, it would spontaneously relax into that optimum. But such arrangements are rare compared to those that can only relax into relatively poor optima.

Watson's research group mathematically simulated the effect of randomizing the initial positions of the anchors and then allowing the system to settle. Over a thousand repetitions, the achieved optima hovered around a level that the next phase of the experiment would show to be unimpressive.

Researchers then modified the simulation by adding a form of self-modeling. Just as in the room-switching scientist experiment, they observed that this small experimental modification permitted the system to bypass poor-quality optima and achieve better ones.

In the scientist simulation, researchers produced self-modeling of past room cohabitation by calculating familiarity factors. For the anchor-and-spring simulation, they turned to a somewhat physically realistic process that produces much the same effect. When a real spring is stretched repeatedly, it permanently elongates; it becomes "stretched out." (The same thing happens with repeated spring compression, but for narrative simplicity, I will leave that element out.) Stretch deformation serves as a form of self-modeling because the extent of deformation tracks how frequently a spring has been temporarily stretched. Furthermore, because it may take hundreds of stretches to cause a spring to permanently deform to a significant extent, the timescale of this memory effect is much longer than the timescale of immediate optimization; it is the equivalent of a Timescale Three process.

In the simulation, a small stretch-deformation factor was added to each virtual spring's future behavior based on how much strain it was under each time the randomized system came to rest. The immediate effect of this alteration was to make each re-randomized system find the usual poor optima a little faster because stretched-out springs are more "willing" to be stretched. One might say that they had come to accept that they would frequently be called on to compromise. In the language of attractors, the basin of attraction for their stretched state had enlarged.

However, something more interesting happened when the process was repeated many times: Accumulating stretch-deformation factors progressively caused the system to land on *better* final optima—those with lower stress-and-strain sums. Importantly, these good optima were also good optima in the non-stretched spring case; the inclusion of stretch-deformation factors caused the system to find the optimum of the original system rather than creating new ones. (This has a parallel in the room-switching scientist simulation where the final scores calculated for each room-switching cycle were based on compatibility factors alone rather than compatibility plus familiarity.) And, once more, the new, improved optima were unlikely to have ever appeared as randomly achieved optima during the prior process of training; accumulated deformation permitted the system to find improved optima that it had never previously sampled.

Thus, this simulated but (somewhat) physically realistic system of anchors and springs spontaneously generalizes from past optimization experiences to find new and improved optimization states. The system "gives way" in response to stress, and that accommodation makes it better able to relieve stress; as the two processes alternate, the system learns to optimize more effectively.

Natural selection operates like the relaxation of the spring system, rapidly climbing the nearest fitness peak. On a slower timescale, the genotype-phenotype mapping gives way to the regularities of the external world by forming Hebbian associations. A do-si-do of the two processes makes future optimization through natural selection more efficient.

Long-term dynamics similar to these may occur in many other systems that possess characteristics equivalent to "giving way." Watson and his group offer simulations that identify similar processes in ecosystems, in which stable patterns of interactions between compatible species become incarnated as Hebbian associations between them. They also show that processes of the same kind may operate in the classic major evolutionary transitions where collections of independently reproducing unicellular organisms transform into jointly reproducing multicellular wholes. (See Watson et al. 2016 for more detailed descriptions of these "evo-eco" and "evo-ego" processes. See also his developing ideas of "natural cognition" presented in the *Songs of Life and Mind* series.[68]) And it doesn't end there; they have also shown that Hebbian processes operating on a local level can cause simulated societies to achieve global economic optima.[69]

Indeed, once one starts looking for it, the same or similar phenomena begin to reveal themselves in unexpected places.

In the discussion of parameterized bird beaks above, I noted that the recurrent need for beak shape alteration in the history of bird evolution has yielded an efficient system for altering beak shape through the adjustment of two or three developmental dials. Developmental parameterization is a kind of self-modeling, not of static states but of repeatedly traveled paths through adaptation space. This process loosely resembles the way that footpaths become more deeply impressed the more often they are used, a fact that hikers, deer and—in a related way—ants exploit.

Could related analyses provide insights into mysterious optimization processes that occurred so early in life's origins that fully Darwinian processes did not yet exist? One such example is the emergence of a single genetic code, a system so optimized that it would seem to have (circularly) required something very much like itself to come into being. And what about stages even earlier than that? According to the metabolism-first theory of the origin of life that I describe in my book *Spontaneous Order and the Origin of Life* (based on the ideas of Eric Smith and Harold Morowitz), pre-life organic chemistry gradually ordered itself into a form that maximizes energy flow. Smith and Morowitz analyze this process in terms of autocatalysis; chemicals that even marginally catalyze their own synthesis would tend to dominate early Earth chemical systems. However, the way they work out this concept in detail involves elements of modularity and self-modeling that resonate with Watson's theories.

The possibilities are tantalizing. Natural induction through self-modeling may be a breakthrough paradigm.

More Detailed Intuitive Explanation

The simulations performed by Watson and his associates show that simple forms of memory can allow systems to generalize. As Watson stresses, there should be nothing surprising about this capacity: models built by sampling a broad subset of possibilities often work beyond their training sets. That is why artificial neural networks trained on images of cats found online can identify cat images they have never before encountered, why ChatGPT can compose responsive, coherent, non-plagiarized texts and why deer can run effectively through novel terrain.

Nonetheless, the idea that mutations operating through developmental systems can use inductive generalization to "suggest" appropriate adaptations tends to strike people as implausible. In this subsection, I attempt to offer a second intuitive explanation based on the recombination of macroscopic adaptations, although in doing so, I will soon run into the limits of verbal explanation.

Let us suppose that deer-sized herbivores can evolve to fill one of two niches: The first requires a caribou-like ability to migrate long distances, while the second requires advanced mountain goat-like skills at leaping up steep rocky slopes. The caribou lifestyle requires long legs and a predominance of slow-twitch, endurance-adapted muscle fibers; the mountain goat lifestyle depends on short legs and fast-twitch, strength-adapted fibers. Let us further suppose that fast-twitch muscles combine poorly with long legs and slow-twitch muscles combine poorly with short legs.

Without Hebbian connections, it would be difficult for an herbivore to evolve from the slow-twitch, long-legged condition to a fast-twitch, short-legged condition because each element of these paired changes is harmful alone, and only in the rare cases that both happen simultaneously would the alteration produce a fitness benefit. But if each mutation has a one in a million chance of occurring, the chance of both happening at once is one in a trillion; thus, the mutational distance between the two states is very great.

Let us suppose, however, that in the evolutionary history of the organism, it has evolved through a complex series of stages that have frequently induced the short-legged, fast-twitch condition and, at other times, have induced the long-legged, slow-twitch state. As a result, strong, redundant Hebbian connections will have emerged that link short-leggedness and fast-twitch musculature and long-leggedness and slow-twitch musculature. If a population of long-legged, slow-twitch deer encounters environmental conditions for which an ability to leap offers survival benefits, a single mutation—one that produces either short legs or fast-twitch muscles—will suffice to achieve the transition. Existing Hebbian linkages will bring the associated change along automatically.

Note that over the course of evolutionary time, many other changes might have occurred in these deer. The switch from the short-legged, fast-twitch muscle state to the long-legged, slow-twitch state is an adaptation along a single dimension; what will emerge each time it happens may be a novel organism unlike any previous one, but that has made a parallel evolutionary move in certain elements of its physical structure. Furthermore, the environmental conditions that induce the shift need not exactly match the ones that caused past organisms to make the same change; they must merely be similar enough in some essential ways that the switch is adaptive. In the language of generalization, the switch between these leg traits generalizes over various deer bodies and various environmental conditions.

But so far, this is only developmental/evolutionary memory. To show the more subtle generalizing capacities of Hebbian association, we need to add more traits.

A lot of traits. So many that the following three paragraphs are almost incomprehensible and, even so, don't quite capture it. (They do make sense if read closely, but feel free to skim them.)

One might suppose, for example, that deer can have a long body suitable for thinly wooded environments or a short body well adapted for movement in densely wooded ones. In the evolutionary history of these imaginary deer, only long-bodied deer living in thin forests have ever had to leap a lot (and, therefore, had short legs and fast-twitch muscles); the combination of short bodies, dense forests and conditions requiring high leaps has never occurred. Now, suddenly, conditions change such that a population of short-bodied, long-legged, slow-twitch-muscled deer encounters an environment that its ancestors never faced: dense woods where the ability to leap is preferable to running long distances but short bodies are more fit than long bodies. As a result, evolution asks for something novel: a short-bodied deer with short legs and fast-twitch muscles.

Hebbian connections make this easier. Any mutation that shortens legs will also bring about a shift to fast-twitch muscles, and vice versa. Thus, the mutational distance between short-bodied, long-legged, slow-twitch deer and short-bodied, short-legged, fast-twitch deer is reduced; an entirely novel phenotype, one that no deer ancestor has ever possessed, stands close to the current phenotype in mutation space.

However, there's a problem: One has to expect that the three conditions of long body, long legs and slow-twitch muscles would themselves have become linked in a Hebbian fashion, and this connection would tend to restore the triad if any one element changed. One can get around this obstacle by adding additional traits linked by Hebbian relationships of varying strengths, and enough of them so that what occurs can illustrate generalizations more complex than those of developmental memory. Interactions sufficiently complex to be interesting cannot easily be described in trait-by-trait detail. Instead, they must be computed.

Kostas Kouvaris and his associates performed simulations that examined the behavior of evolving systems with many traits and trait relationships (Kouvaris et al. 2017). The results of those simulations showed that Hebbian associations can indeed steer trait combinations away from poor local fitness peaks and toward more optimal ones, including trait combinations that had never previously been sampled.

Furthermore, when the patterns that emerged are examined closely, they can be understood as generalizations.

To put this in the language of springs and anchors or room-switching scientists, the Hebbian connections between traits serve as memories of local elements of past optima. Those memories not only allow future evolution to reproduce those local elements more efficiently but also permit those systems to find new global optima that were never achieved before.

For readers familiar with neural network behavior, certain additional wrinkles in the research results will be immediately recognizable. I will merely mention them here and recommend the full paper to those interested in the details. (Like most of the other papers listed in the Major References section, Kouvaris et al. 2017 is available in free full text. As an aside, its title, "How Evolution Learns to Generalise," is so perfect that I was dangerously tempted to steal it.)

The system studied by Kouvaris and his coauthors proved highly sensitive to the rate of environmental switching. When environmental conditions were set to change too slowly or quickly, generalizing capacities did not develop; the behavior of the system at those extremes resembled the well-known neural network issues of overfitting and underfitting, respectively. However, when the researchers introduced noise into the environmental data and also included incentives for parsimonious Hebbian connections, the model became more robust. Since actual environments are noisy and there is plausibly a cost to excessive Hebbian complexity, these tweaks may be biologically realistic. (How these real-world factors spontaneously find a Goldilocks balance between overfitting and underfitting remains to be explained.)

These research results are remarkable, but a by-now familiar major caveat applies: Just as with the other simulations described above, Kouvaris et al. only modeled pairwise Hebbian relationships of idealized traits and not the nested, hierarchical relationships that operate in actual GRNs (and in brains and artificial neural networks). Again, this is a significant limitation. Nonetheless, analogy with the learning processes of deep-learning artificial neural network systems suggests that the same processes should continue to operate but on an increasingly subtle level. Just like large language models learn to identify deep patterns in text and can utilize that knowledge to analyze, process and produce novel text, sufficiently complex hierarchical GRN Hebbian structures should be able to identify subtle patterns of Earth environments and "use" that knowledge to supply appropriate adaptations when confronted by novel conditions.

If this analogy is correct, the network structure of the GRNs that operate in development will have been shaped by nested Hebbian associations in such a way that the near neighbors of any given phenotype—the ones that can be reached by simple ridge crossing—in a certain sense, *represent* variations responsive to particular ways that environments have been known to shift in the past. For example, if Earth condition A periodically changes to Earth condition A', and does so independently of all other conditions, then some neighbor or neighbors of the current phenotype will alter an adaptation for A into an adaptation more suited to A' while leaving other aspects of the phenotype largely unchanged. If, however, the change from Earth condition A to A' is typically associated with a simultaneous change from condition B to B', there will be a ridge that, once crossed, shifts the adaptations to both at the same time by modifying the operation of a higher-level module that controls the two as sub-adaptations. Still more complex Hebbian interactions will result in still more subtle adaptations.

The way this works is easiest to understand at the level of an independently varying A to A' shift without Hebbian connections, and where the responsive phenotypic alteration can be accomplished through the manipulation of a single, continuously variable trait. For example, suppose that the favorite prey of a predator begins to trend larger under selection because that change reduces its vulnerability; this is a common evolutionary "strategy" that has been utilized by numerous prey animal lineages in numerous conditions. Predator species typically respond by evolving larger sizes, too. The environmental change from

A to A' is a shift in prey size, and because that shift has occurred so often in the past, predators also possess a GRN dial that easily adjusts their size. The correct adaptation to the change in conditions is thus readily available and can be accomplished without altering any other trait.* In the language of springs and anchors, the basin of attraction for a shift in body size has been enlarged through multiple past experiences in which that change occurred.

Once more, the A to A' shift in the environment doesn't have to exactly mirror any prior change; it only needs to possess some sort of structural similarity to a past change. An increase in size of *any* prey species implicates much the same predator adaptation, whether the prey is a dinosaur or a deer-like mammal. To put this another way, change in total body size is a kind of skeleton-key evolutionary transformation that generalizes over many challenges.

However, while certain dials can rotate freely, the types of phenotypic changes that they control are constrained and guided by downstream Hebbian connections. A deer-like creature with four long legs lies close in mutation space to one with four short legs; perhaps a single mutation affecting a single continuous dial will do it. However, a deer whose left front leg alone is lengthened lies far away in mutation space from the original deer, and appropriately so, because that combination of traits is unlikely to be adaptive.

The neighborhood relationships of the features and sub-features utilized by GRNs to produce different phenotypes resemble the neighborhood relationships of data features that neural network image recognition systems attain as they train on datasets. Suppose an image-recognizing program is presented with three images of a deer, the first with four short legs, the second with four long legs, and the third with three short legs and one long leg. The deer with four short

---

* The change in size of the predator now acts as an A to A' shift in the environment of the prey, and it too will respond. Presumably, it was a do-si-do of this kind that caused both tyrannosaurs and titanosaurs to evolve their enormous bodies.

legs is closer in "image representation space" to the one with four long legs than to the one with three short legs and one long leg. Image-recognition programs learn to place related images near one another by training on numerous images with recurrent patterns; GRNs learn to place reasonable adaptions nearby in mutation space by evolving through numerous Earth conditions that also have recurrent patterns.

This arrangement limits the number of phenotypic changes that can occur via single regulatory mutations, but there are still many possibilities. Most will prove non-adaptive; or, rather, adaptive to different environmental conditions than those that are present. Organisms with those mutations will constantly appear, but selection will remove them.

The geometry of phenotypic neighborhoods may have an additional important property. Suppose that to achieve a successful adaption, the GRNs of an organism need to undergo a sequence of ridge-crossing events (i.e., A to A' to A'', etc.). If some of the necessary intermediate changes cause a temporary loss of fitness, the sequence won't be achievable under selection. However, in the flat, pairwise simulations described above, familiarity factors create pathways toward better optimization in which each step is favored. If the same holds for systems with nested, hierarchical Hebbian relationships, there will typically exist a sequence of steps toward an effective adaptation where each increases fitness, or at least doesn't decrease it.

Scales and Evolutionary Moments

In the preceding section, I illustrated GRN Hebbian generalization in terms of macroscopic traits, such as total body size, leg length and body length. However, while it is easiest to discuss such large, macroscopic changes, Hebbian generalization seems unlikely to have its most significant influence at that level. Mechanisms like heterotopy, heterochrony and various forms of parameterization may be the source of most changes in visible form, and these simple, top-level modifications do not give much scope for non-trivial Hebbian generalization.

In the preceding sections, I was forced to abandon verbal descriptions and turn to the Kouvaris et al. simulation because interesting generalizations only occur when there are many traits and trait relationships. This suggests that non-trivial Hebbian generalization may occur most often in the recombination of low- to medium-scale traits (sub-sub-traits and sub-sub-sub-traits, etc.) simply because there are so many more of them. For example, perhaps linkages exist between claw length and the detailed chemical composition of claws so that whenever claws are induced to grow longer, their elasticity, strength, growth rate, shape and resistance to wear also change adaptively.

Once future scientists have mapped out a comprehensive, fine-grained picture of GRNs and their changes over evolutionary time, they might observe a constant, subtle recombination and adjustment of traits beneath the easily observed macroscopic level. Watson's theory predicts that non-trivial generalizations will occur in the relationships between these traits as they relax into higher levels of fitness. Importantly, this prediction provides direction for future research by suggesting that scientists consciously *look* for such effects.

Contrary to popular belief, science only sometimes proceeds through brute force data collection; more often, researchers begin with a theoretical perspective that tells them what characteristics of the data they should focus on.

However, at the current state of knowledge, information about trait combinations at that level is sparse to non-existent. So little is known that I can't even invent a totally hypothetical example. Only as data gradually comes in will it become possible to discuss the theory in the same rich biological detail as the other aspects of facilitated variation.

Hebbian generalization might also play a particularly significant adaptation-enhancing role during major evolutionary transitions such as the ancient emergence of multicellularity and, later, of the animal toolkit. During such transitions, not only regulatory elements but also housekeeping genes (the lowest-level modules) undergo large changes. The processes that led to major evolutionary transitions have been difficult to explain with traditional evolutionary models.

For example, the universal toolkit for constructing animal bodies is so sophisticated and flexible that it must have undergone a lengthy optimization process to achieve its final form, and yet it seems unlikely that this optimization process was limited to Darwinian selection alone for the usual reason that selection can only favor current abilities, not future flexibility. Nonetheless, as the anchor-and-spring simulation showed, selection isn't necessarily the only optimizing force at work in evolution. Timescale Three adaptation through natural induction may have combined with Timescale Two optimization through natural selection as they together brought the animal toolkit into existence and refined its capacities.

Hebbian Generalization vs. Other Forms of Facilitated Variation

Watson's work offers GRN Hebbian generalization as a novel mechanism of facilitated variation, one that allows evolutionary processes to avoid getting trapped on poor fitness peaks and efficiently find better ones. If one accepts the arguments for the existence of the mechanism, one might still wonder how much of a contribution it makes compared to the other, better-established factors. However, a bit of reflection suggests that Hebbian generalization is so closely intertwined with the other forms of facilitated variation as to make such comparisons meaningless.

Consider the power of multiuse widgets to assist adaptation. Whenever environments change, there is probably some widget somewhere that can make a helpful contribution simply by being switched on, adjusted or recombined. If deer-like creatures are forced to adapt to terrain composed largely of sharp rocks, their GRNs likely already possess suitable biochemical processes to build tough hooves. Nonetheless, since all widgets above the level of single genes are composed of sub-widgets composed of sub-sub-widgets, the Hebbian relationships that structure these hierarchical modular relationships will play a significant role in biasing how they are activated and modified.

Similarly, it seems plausible that among organisms that have evolved sophisticated brains, behavioral phenotypic plasticity and the Baldwin effect may play a considerable role in guiding variation. Since all but the simplest brains are much more sophisticated than GRNs, the Hebbian generalizations that matter most to animal evolution may occur in actual neural networks rather than in the neural network-like GRNs that control development. Nonetheless, even when adaptations are guided by behavioral phenotypic plasticity, the genetic accommodation that follows will proceed along the lines most easily accessible within the existing pattern of Hebbian connections and parameterized pathways. Here, as elsewhere, various forms of optimization optimize one another, and logically distinct forms of generalization increase each other's generalizing powers.

Is Watson's Theory True?

The principles of facilitated variation originally promulgated by Kirschner and Gerhart are all now well established. The ability of phenotypic plasticity to guide evolution has been extensively documented by West-Eberhard and others; no one doubts that exploratory and adaptive processes reduce the number of mutations necessary to induce changes in animal forms; it is not controversial that the use of other general-purpose tools, parts and features make construction more efficient and that modular organization combined with weak linkage offers a parallel benefit at the level of design.

To this set of ideas, Watson adds three more based on Hebbian relationships in GRNs. To recap, they are as follows:

- Hebbian relationships produce and maintain modularity.
- Hebbian relationships produce and maintain developmental/evolutionary memories.

- Generalization based on Hebbian modeling of the world causes responsive, educated guesses—plausibly adaptive variations—to lie only a relatively short sequence of hill-climbing mutations away from those of the current phenotype, thus permitting them to be preferentially expressed and tested.

The idea that Hebbian associations contribute to modularity seems to be nearly obvious once it has been pointed out.

The fact of simple developmental memory is proven in atavisms, and the story of snakes that regained partially operational legs shows that damaged modules can be restored, at least at the top levels that form macroscopic traits. However, this evidence only scratches the surface of Watson's theory of developmental memory, which predicts more subtle effects, such as the spontaneous recall of the optimal limb length and distance ratios appropriate for large and small deer-family creatures. It might be possible to marshal fossil evidence that shows the recurrence of ratios, but it would be difficult to detect whether they took on those recurrent values through memory effects or via hill-climbing processes that simply rediscovered the same optima.

It is even more challenging to find direct biological evidence for Hebbian generalization. Our current knowledge of the low-and mid-level trait relationships where Hebbian generalization plausibly plays its greatest role remains far too coarse and gappy to confirm (or deny) the theory.

Thus, computer simulations remain the only meaningful support for the proposal that Hebbian structures in GRNs support subtle forms of developmental memory and perform interesting generalizations. However, as mentioned several times above, the simulations performed by Watson and his group only model systems with flat, pairwise Hebbian connections rather than the nested, hierarchical structures of actual GRNs. Such simulations are inherently limited in their ability to model *context*.

To see what this means, let us consider analogous processes in artificial intelligence.

Large language models such as ChatGPT operate by predicting what word should follow any given word. The simplest way to accomplish this would be to study words in pairs. Suppose that, in all texts available online, the word "married" is most frequently followed by the word "couples." If ChatGPT only modeled language at the level of pairwise association, it would always follow "married" with "couples" and never with "to," "in," "late," or any of the other possibilities. Pairwise modeling is a form of contextual learning, but it is too simplistic to model human language well.

ChatGPT achieves its fluency by going beyond pairwise word relationships and studying the detailed patterns of *neighboring* words. The probability that one word will follow another depends on the surrounding text. If a paragraph includes the words "guests," "views of the ocean," and "the bride was nearly drowned by a sneaker wave," the word "married" is more likely to be followed by "in a ceremony by the beach" than by "couples." Conversely, if adjacent text contains "single parents" and "evolving characteristics of family units in US society," "married" is indeed likely to be followed by "couples." ChatGPT produces comprehensible and responsive texts because it models context deeply.

Something similar can be found in the GRNs that build organisms. It may be the case that when trait X is expressed, trait A is Hebbian-linked to trait B, but when trait Y is expressed, trait A is Hebbian-linked to C and not to B. Such a differential linkage could be instantiated in the form of two modules: one that calls X, A and B together and another that calls Y, A and C. Processes like these are necessary in systems that construct organisms by using the same sub-traits as components of various higher-level traits. For example, the modules that build bone are utilized in numerous locations and do so in association with other modules, but those associations differ depending on where the bone-building module is activated. The same is true, but even more profoundly, at the level of genes, whose pleiotropic expression spans numerous distinct modules and processes.

However, flat simulations limited to pairwise Hebbian associations are fundamentally incapable of modeling the deep context-sensitive processes so central to developmental processes. In the parlance of mathematical physicists and economists, these are "toy models," consciously simplistic representations of systems that are difficult to model fully.

There is nothing wrong with using toy models to grapple with complex problems. Science does so frequently. The Ising model of ferromagnetism, the lattice model of quantum field theory, the IS-LM macroeconomic model used by Keynesian economists and the DeSitter space studied by cosmologists are all toy models, and each has made important contributions.

In a deeper sense, *all* useful models are toy models in that they depend on numerous simplifying assumptions. It would not be entirely wrong to say that scientific genius consists largely of the ability to identify and exploit useful simplifications. However, finding useful simplifications in the field of complex systems, of which life is the prime and most challenging example, has proved difficult. Any toy model that captures elements of the behavior of complex systems would be a welcome contribution.

Classic Darwinian selection, as represented in the current "orthodox" view of evolution, the modern evolutionary synthesis, is itself a highly simplified model that has proven extremely useful for understanding evolution; its many offshoots (such as population genetics) do a good job of explaining certain evolutionary processes. However, the Darwinian model in effect assumes that useful variations fall from heaven; it does not attempt to explain where variations come from or what forms they are likely to take. Furthermore, the standard Darwinian model has difficulty grappling with evolutionary self-action, such as the change in level of individuality that accompanied the emergence of permanent multicellular organisms or the two-way interaction of organisms and their environments.

Economies, too, are complex systems, and attempts to model economic behavior have similarly run into difficulties. All economic models are, of necessity, highly simplified and manifestly imperfect; nonetheless, experience has shown that widely-used macroeconomic models of such phenomena as trade flows, inflation, spontaneous price-setting and elastic vs. inelastic demand are significantly more predictive than coin-flipping and undoubtedly capture certain fundamental elements of economic activity. Furthermore, these imperfect, simplified models are amenable to incremental improvements that improve the match between model and reality. Nonetheless, they sometimes make incorrect predictions.

We can expect a deep understanding of evolution to be at least as difficult as the challenge of predicting economic processes. One has to start somewhere and go from there. Watson offers an initial step.*

Scientists who break new ground often find it necessary to begin by adopting the paradigm that they hope to prove because it is only by looking through the lens of a correctly chosen paradigm that they can spot the regularities predicted by that paradigm—if any exist. Copernicus had the intuition that the Earth revolves around the Sun, but his proposed system of circular orbits yielded worse predictions than the existing geocentric system of Ptolemaic epicycles. Kepler kept heliocentrism but abandoned the perfect circles and built his model on ellipses. Without Copernicus's paradigm, Kepler wouldn't have known where to start (and, even so, what he accomplished was remarkable). Watson and his group invite us to look at evolutionary processes through the paradigm of natural induction and see what we can discover; they suggest that once we possess detailed knowledge of traits at a fine-grained level, we should look for evidence of subtle developmental memory and broad-spectrum generalizing processes.

One core strength of Watson's hypothesis is its specificity. Evolution has long been compared to a learning process, but Watson goes beyond vague analogies and spotlights the mechanisms of learning known to occur in artificial and natural neural networks. Such specificity is

---

* If you, my reader, have contacts in the world of computer science foundations, perhaps you can find someone to offer Watson and his group the expensive computational power necessary to go beyond pairwise Hebbian relationships!

important because it can elevate metaphor into equivalence. It was Einstein's recognition that acceleration and gravity are not just similar but (locally) indistinguishable that led him toward general relativity. If evolution involves processes that closely map onto neural network learning, researchers may be able to acquire deep insights into evolutionary processes by studying neural networks, and the latter may in some ways be more tractable than the former.

Good models can explain facts about the world that were previously confusing and predict the discovery of new facts that were otherwise unsuspected. Einstein himself showed that general relativity could explain previously mysterious anomalies in the orbit of Mercury, and Karl Schwarzschild used the mathematics of general relativity to predict the previously unsuspected phenomenon of black holes.[†] If evolution can be successfully modeled as a neural network learning system, simulations based on that correspondence may explain currently confusing facts of evolution as well as predict previously unrecognized evolutionary phenomena.

Eventually, the theory of natural induction through self-modeling will require some sort of experimental confirmation, but that doesn't necessarily have to happen soon. Darwin's theory of evolution through natural selection was a breakthrough long before there was any way to test it experimentally; plausibly, it wasn't until the late 20th century and the advent of gene sequencing that Darwinian selection began to move from "logically persuasive and consistent with many observations" to "supported by experimental evidence."

Nonetheless, experimental confirmation of Watson's theories would be useful.

It might not be too difficult to study his more modest claim that Hebbian connections tend to preserve developmental/evolutionary

---

[†] Schwarzschild himself didn't fully recognize the implications of what he'd discovered. The modern idea of black holes took shape over subsequent decades and included contributions from many scientists, but most of it is already present in Schwarzschild's original paper.

memory and can help restore memories that have been damaged. Here is a schematic outline of how it might be done, following suggestions by Tobias Uller.[70]

First, choose a rapidly reproducing organism and divide it into two populations. Subject one of these to novel condition A and the other to novel condition B. Allow each population to adapt. Next, remove conditions A and B and apply condition C to both groups. Give both populations time to hill-climb into a new stable state. Finally, switch out condition C for condition A or B and measure the time it takes for the organism to adapt to its ancestral versus non-ancestral environment, using new mutations.* According to theory, readaptation to an ancestral environment (i.e., sequence A-C-A and B-C-B) should, on average, occur more quickly than initial adaptation and quicker than adaptation to environments not experienced in the past (i.e., A-C-B and B-C-A). Follow-up experiments could investigate how rapidly and by what evolutionary pathways adaptations are "rediscovered" when condition A or B is restored.

Hebbian generalization presents a greater challenge. Deep knowledge of patterns in low- and medium-level traits might provide *observational* evidence of Hebbian generalization, and that alone would go a long way toward supporting Watson's theory. However, it seems to be conceptually difficult to design *experiments* that can identify forms of Hebbian generalization.

One might naively consider comparing the rate of evolution in organisms whose GRNs include Hebbian associations against others whose GRNs had not been so shaped. Alas, according to the theory itself, all eukaryotic organisms already model the world to a high degree, and

---

* In his review of a version of this text, Uller mentioned a tricky subtlety: the need to control for standing genetic variation. It is always possible that low frequency, hidden (recessive) allelic variants might be responsible for rapid adaptations. Since this research aims to find characteristics of the genotype-to-phenotype map that make reversals more likely, experiments need to be carefully designed to eliminate that possibility, perhaps by seeding populations that encounter new selective regimes with a single genotype or letting them pass through a severe bottleneck.

even bacteria possess Hebbian shaping in the form of operons. Thus, there are no non-Hebbian organisms on which to try such experiments.

Despite these difficulties, I would not in any way rule out the possibility that someone will devise a way to test the theory in the laboratory. The history of science is full of brilliant experimenters.

Implications for Global Climate Change

> *"What was, will be again, what has been done, will be done again, and there is nothing new under the sun."*
> ECCLESIASTES 1:9 [71]

Ocean acidification and rising water temperatures are currently stressing coral reefs by impairing coral skeleton formation and killing the symbiotic algae upon which corals depend. This has raised serious concerns about coral extinction. However, even leaving aside Hebbian generalization, the ideas presented in this essay suggest that corals as a whole will probably do just fine, although the same is not necessarily true of the many other organisms that depend upon them.

Corals first emerged almost half a billion years ago and have persisted without too much change through multiple warming and cooling cycles, accompanied by changes in the levels of dissolved carbon dioxide and ocean acidity. Because of this history, corals and their symbiotic algae living today likely possess a greater degree of phenotypic plasticity than we currently recognize.[72] They also likely have access through developmental/evolutionary memory to adaptations that their ancestors employed to survive in warmer, more acidic environments.

Since the first appearance of corals in the fossil record, Earth's average temperature has varied widely. During the Cretaceous Thermal Maximum, 85–90 million years ago, the average ocean temperature reached levels perhaps 5–7 degrees Celsius (9–13 degrees Fahrenheit) above recent pre-industrial levels, significantly surpassing all but the most extreme and (probably) unrealistic forecasts of the consequences of anthropogenic global warming.[73] Since the fossil record shows evidence of significant coral populations throughout that period, they either already possessed or rapidly evolved adaptations that allowed them to survive those warm conditions.

When temperatures fell after that, one can safely assume that the corals initially responded through their inherent phenotypic plasticity, but if cooling went past the limits of their plastic responses, their GRNs and genes would have come under selection for survival in cool environments. Based on the principle of evolutionary memory, the most accessible adaptations would have been those that the corals had used prior to the rise in temperature, and these would have been rapidly restored, although perhaps a bit of simple hill-climbing would have been necessary to repair any adaptations that had begun to decay.

At around 55 million years ago, Earth once again experienced high temperatures that were sustained for approximately 200,000 years before conditions cooled again. In response to this temperature rise, corals would have brought out their earlier high-temperature adaptations and, in so doing, refreshed their evolutionary memory of them.

It is often assumed that the current change in global climate is happening much faster than the past changes in climate that can be seen in the fossil record. However, this may not be the case. Rapid fluctuations quickly become invisible when one looks into the distant past. If, 30 million years ago, temperatures spiked and then fell again over periods as long as 10,000 years, we wouldn't be able to detect them. It's much easier to detect rapid changes in more recent times. And, considering what is known of the last several million years of climate history, extraordinarily rapid climate changes may be common.

During the recurrent ice ages of the Quaternary period, beginning 2.58 million years ago, local and global climate underwent repeated swings so rapid that the entire period has been characterized as one of climate chaos. Average temperature shifts of 5–10 degrees Celsius over

50-year periods occurred frequently. Rainfall, too, underwent sudden changes that caused large lakes to dry out, fill to bursting and dry out again on similar timeframes.[74] Models of the effects of anthropogenic climate change do not predict changes any more dramatic than these. Were temperature fluctuations as rapid during earlier periods of rising temperature as they were in this recent period of falling temperature? We have no way of knowing, but it seems plausible.

Another factor corals have going for them is their immense current biomass. Natural variation will allow some current corals to survive better than others, not only due to genetic differences but also to accidents of location. As corals face high levels of selection pressure (a euphemism for "mass die-off"), some will tolerate it better than others. It would not be surprising if more than 90 percent of all corals die, but that would leave an enormous population that adapts, survives and, after a delay of a century or two, bounces back.

The promise of future coral recovery will not, however, help the many species of marine animals and plants that depend on coral reefs (although they, too, may adapt in unexpected ways). Furthermore, selective processes might favor heat-tolerant coral species that do not provide a suitable habitat for other organisms—a process that may be occurring already.[75]

Land animals and plants may be even better situated. Almost all species alive today survived ice age climate chaos while a great many others went extinct. Evidence suggests that mammalian species with larger brains were less likely to go extinct, perhaps because their greater braininess equipped them with an increased capacity to adapt behaviorally as conditions shifted.[76]

As we now enter a period of anthropogenic climate change in which temperatures stand a good chance of rising as much as 5 degrees Celsius (9 degrees Fahrenheit) over the next hundred years, these prior adaptations will be tested. We will likely discover a great deal of adaptability that has been hidden for a while. The Earth's climate has been unusually stable for about the last 12,000 years, a period known as the Holocene.

Of course, the unstable climatic conditions of the previous few million years varied around a lower mean temperature than what is coming, and many forms of plasticity useful for surviving ice age conditions won't work in higher temperatures. The ability of animals and plants to rapidly evolve adaptations in the face of anthropogenic climate change may depend on the experiences of even more distant ancestors that managed to thrive during periods when average temperatures were equally high.

Nonetheless, neither phenotypic plasticity nor evolutionary memory is a panacea, and many species will certainly go extinct when the climate warms. The challenge of climate change will be made more difficult by other aspects of the massive ecological footprint of eight billion humans and their massive energy use per capita;* fish already stressed by over-fishing and animals whose habitat has been degraded or constricted by human development may not have the time or sufficient population to adapt to changing climate. We will lose many plants and animals that matter a great deal to us, and we humans will undoubtedly suffer terribly. But the advanced generalizing powers possessed by current organisms, including ourselves, will likely allow the living world to do better than some grim forecasts predict. Diversity will return, if in different forms. (Hopefully, this will not be limited to hundreds of species of rats and thousands of species of cockroaches.)

Epilogue: The Evolution of Evolvability

I originally planned to give this essay the title "The Evolution of Evolvability" because it has a nice ring to it and conveys the central idea neatly. But I refrained because doing so would be misleading on a central point.

---

* This is not to suggest that humans are evil for creating such an impact. Highly successful species typically induce strong selection pressure on other organisms, and the exhausting of food supplies due to over-predation is the most common factor that limits predator populations. We evolved under selection to become a powerful selective force.

The general notion of evolution preceded Darwin and originally implied nothing beyond gradual change. Darwin's particular contribution was to offer a mechanism for evolution, that of natural selection on heritable variation. Nonetheless, the older meaning of evolution continues to be used in popular culture. Technology, Pokémon and certain superheroes are said to "evolve," in the sense that they change and advance. Given this colloquial usage of the term, evolvability can be said to have evolved, too.

But Darwin's contribution permanently modified the commonly understood meaning of the term "evolution" to include selection on heritable variation, and, as has been discussed several times above, there is no obvious way that a population of organisms can come under direct selection for evolvability.

One could naively try to claim that evolvable organisms have more descendants than less evolvable ones and, thus, they come to dominate over long periods. However, according to one (admittedly crude) estimate, about 80 percent of the planet's biomass consists of plants, 15 percent are bacteria, 2 percent are fungi, and 0.3 percent are animals, with archaea, viruses and unicellular eukaryotes filling out the remaining 2.7 percent.[77] If animals are the paradigmatic examples of highly evolvable organisms, it doesn't show in these bulk numbers.

Nonetheless, in the final chapter of *The Plausibility of Life*, Kirschner and Gerhart make an apparently half-hearted attempt to show that a form of meta-selection might have contributed to the origin of facilitated variation. I present it here not because it works out well but to show that explanations based on natural induction through self-modeling fill a void.

In brief, their argument goes as follows: When rapid changes in Earth conditions drive a high percentage of current species to extinction or otherwise provide many new opportunities, the most evolvable organisms are the ones most likely to rapidly diverge genetically into newly available niches. The term for this process in evolutionary biology is "radiation." Like the branches of a tree numerous new species emerge from the trunk, the branches of the trunk and the branches' branches. A massive radiation of mammals occurred after the dinosaurs went extinct; something similar happened later with birds. Conceivably, this occurred because mammals and birds, or at least those of them that gave rise to many species, were particularly evolvable.

Going back much further, the Cambrian Explosion produced a truly massive radiation of animal forms, perhaps because the emergence of the modular embryonic toolkit used by animals enhanced their evolvability. Radiation events might act as a form of meta-evolution that rewards the most evolvable species. In that case, those current species and groups of species found in the highest number of niches would be the descendants of highly evolvable species and, thus, may also be highly evolvable themselves.

But this explanation has several problems, not the least of which is the challenge of measuring evolvability. There are approximately 400,000 known species of beetles today. Does this indicate that beetles are more adept at evolving than other insects? Or does it just mean that they easily form reproductive cliques that dispose them to form numerous non-interbreeding species with relatively unimportant differences? It's not really the number of species that is important to evolvability—it's the increased ability to evolve adaptive solutions to novel environmental challenges. But how would one measure that?

Moreover, all eukaryotic organisms seem to utilize every element of facilitated variation; given this, how could meta-selection for evolvability favor certain lineages over others? Would it be favoring those with better implementation of those facilitating factors? Those with better modularity? Widgets of greater general-purposeness? Smarter exploratory processes? It all seems rather vague.

But if we step back from the general subject of the evolution of evolvability and look at its individual elements, the question seems less mysterious.

One of the pillars of facilitated variation, the Baldwin effect, is not particularly difficult to explain. Phenotypic plasticity itself is readily understood as emerging through the ordinary processes of natural selection (or slight variations on that theme in the case of multigenerational plasticity). The fact that phenotypic plasticity can facilitate future evolution is merely an example of the well-known phenomenon that models can often be utilized beyond their training sets; phenotypically plastic responses model changes in environmental conditions that occur over short periods, and these resemble the changes that occur over longer periods. Thus, phenotypic plasticity has the power to focus adaptive evolution along tried and true solutions.

The main puzzle that selection on lineages really needs to solve—but doesn't solve very well—is the origin and maintenance of pluggable modularity. Here, Watson's theory steps in, grounding pluggable modularity in Hebbian relationships that reflect the correlational structure of changing Earth environments. In the language of natural induction, a gradual process of "giving way" to the correlational patterns of Earth environments has progressively altered the genotype-phenotype map in such a way that short-term optimization has become increasingly efficient. Natural induction rather than natural selection is the primary source of pluggable modularity.

Watson's approach unites *all* the elements of facilitated variation by regarding them as forms of generalization across the tasks and challenges that organisms face. General-purpose widgets generalize across tasks; phenotypic plasticity generalizes across short-term environmental shifts; pluggable modularity through Hebbian association generalizes over long-term patterns in how environmental conditions shift; the emergent geometry of phenotypic neighborhoods generalizes about good ways to combine and recombine modules. These various forms of generalization all work together to enhance evolvability because they generalize in different modes and on different timescales.

This collection of independent phenomena grouped together as facilitated variation arises within a complex world of biological detail but feels independent of those details. One senses the presence of deep principles and grand mathematical theorems waiting to be discovered that will describe how and when optimization processes spontaneously arise and iterate on multiple timescales, under what circumstances self-modeling yields powerful forms of generalization and by what other mechanisms evolutionary processes have managed to successfully explore the extremely high-dimensional spaces of possible traits and their relationships.* These are profound questions in the field of complex systems, a science that is still in its infancy, but whose mysteries are as challenging as the quantum world of subatomic particles and—unlike some better-known fields of study, such as string theory—are relevant to many questions of direct human concern. Watson's work may provide penetrating insights into this as-yet barely explored realm.

---

* Related questions arise in machine learning. One of the most fundamental is "the curse of high dimensionality." In the early days of machine learning, it was widely believed that artificial neural networks designed to model high-dimensional spaces such as language and images couldn't possibly succeed because the size of those spaces is too vast. Adding more layers and increasing computational power didn't seem likely to help; the issue is similar to the incommensurable difference between the age of the universe and how long it would take monkeys to randomly type out *Hamlet*—multiply the number of monkeys by a billion or giving them the ability to type a trillion times faster wouldn't make any difference at all. Nonetheless, a few entrepreneur/researchers bullheadedly went ahead adding layer upon layer and employed increasingly powerful cloud computing resources to raise the computational power of their systems. To everyone's surprise, perhaps including their own, their efforts eventually and rather suddenly succeeded. No one has yet been able to explain mathematically why large language models and similar systems are so effective. Could the currently unknown mechanisms and properties that permit artificial neural networks to succeed beyond expectations also help explain the evolution of evolvability?

Major References

Carroll S. *Endless Forms Most Beautiful: The New Science of Evo-Devo.* W.W. Norton and Company; 2006.

Davies AP, Watson RA, Mills R, Buckley CL, Noble J. "If you can't be with the one you love, love the one you're with": how individual habituation of agent interactions improves global utility. *Artif Life.* 2011;17(3):167–181.

Erwin, DH and Valentine JW. *The Cambrian Explosion: The construction of animal biodiversity.* Roberts and Company; 2013.

Kirschner MW, Gerhart JC. *The Plausibility of Life: Resolving Darwin's Dilemma.* Yale University Press; 2005.

Kouvaris K, Clune J, Kounios L, Brede M, Watson RA. How evolution learns to generalise: Using the principles of learning theory to understand the evolution of developmental organisation. *PLoS Comput Biol.* 2017;13(4):e1005358.

Parter M, Kashtan N, Alon U. Facilitated variation: how evolution learns from past environments to generalize to new environments. PLoS Comput Biol. 2008;4(11):e1000206

Uller T, Moczek AP, Watson RA, Brakefield PM, Laland KN. Developmental Bias and Evolution: A Regulatory Network Perspective. *Genetics.* 2018;209(4):949–966.

Wagner GP, Pavlicev M, Cheverud JM. The road to modularity. *Nat Rev Genet.* 2007;8(12):921–931.

Watson RA, Mills R, Buckley CL, et al. Evolutionary Connectionism: Algorithmic Principles Underlying the Evolution of Biological Organisation in Evo-Devo, Evo-Eco and Evolutionary Transitions. *Evol Biol.* 2016;43(4):553–581.

Watson RA, Szathmáry E. How Can Evolution Learn? *Trends Ecol Evol.* 2016;31(2):147–157.

Watson RA, Wagner GP, Pavlicev M, Weinreich DM, Mills R. The evolution of phenotypic correlations and "developmental memory." *Evolution.* 2014;68(4):1124–38.

West-Eberhard M. *Developmental Plasticity and Evolution.* Oxford University Press; 2003.

Glossary

Associative Learning: This is the brain's primary method of learning. If a dog repeatedly hears a bell before being presented with food, it will soon begin to salivate after the bell rings, even if no food follows. The dog's brain circuits that trigger salivation have become connected with those that identify the sound of a bell ringing. They were not wired together at birth but become wired together through experience.

Baldwin Effect: Named after American psychologist James Mark Baldwin, this is the phenomenon by which phenotypic plasticity can guide future evolution.

Cis-Regulatory Element: A considerable portion of the DNA that constitutes a single chromosome consists of regulatory sequences that control the activation or expression of genes on that chromosome. Some of them enhance activation, while others suppress it. In turn, these regulatory sequences can be activated or suppressed.

Darwinian Selection: Natural selection operating on heritable variation.

Eukaryote: An organism whose cells contain internal structures such as a nucleus and mitochondria. Eukaryotes evolved from prokaryotes and are considered more complex or advanced. All multicellular organisms are composed of eukaryotic cells, and there are numerous species of unicellular eukaryotes as well.

Gene: A DNA sequence that codes for a protein.

Gene Control Box: I invented this term to refer to the set of all cis-regulatory sequences that control the expression of a particular gene.

Gene Regulatory Network (GRN): The set of DNA-hosted regulatory sequences that control gene expression. These include cis-regulatory sequences, trans-regulatory sequences (transcription factors) and various regulatory RNAs that are transcribed from DNA but not translated into proteins.

Genotype: The information encoded into an organism's DNA. In the past, this term referred to the set of genes in that organism, but the current proper usage of the term includes regulatory elements as well.

GRN: See gene regulatory network.

Hebbian Association: Neurons that frequently fire at the same time or in close sequence acquire a strong synaptic linkage. This phenomenon is called Hebbian learning, named in honor of Donald Hebb, who first identified it in 1949.[78] Ever since, students of neuroscience have learned the mnemonic phrase "neurons that fire together, wire together." If a single one of the connected neurons is stimulated to fire by outside events, the synaptic connection between them will cause the other to fire, too. The existence of such connections explains why pets respond to the sound of a rustling food package in much the same way as to the smell, taste and sight of the food within it. According to the theories of Richard Watson, gene regulatory networks form their own version of Hebbian associations when they link or anti-link traits.

Hill-Climbing: Algorithms that seek to optimize a parameter through incremental changes are called "hill-climbing" algorithms. Classical Darwinian selection is such an algorithm, and the parameter it optimizes is fitness. The process of evolutionary optimization through mutation is called "climbing a fitness peak," where the peak represents the fittest state available.

Modularity: As this term is used in evolutionary biology, it means that organisms and their actions tend to divide naturally into largely discrete units that interact with one another at only a limited number of points while being tightly ordered internally.

Mutation Space: The set of all possible mutations that can occur in the DNA of a given organism.

Neural Network: A neural network is a set of partially or completed interconnected units that can send signals to one another. In biological nervous systems and brains, the units are nerve cells. In artificial intelligence, neurons are mathematically modeled.

Operon: Proteins are typically useless on their own and can only perform vital actions or create helpful structures when they are produced in the company of several other proteins. To ensure that interdependent proteins are produced simultaneously, prokaryotes collect the genes that code for them into contiguous functional units called operons; when one gene is expressed (converted into a protein), so are all of its "friends." Operons are preceded by switches that can activate or suppress them.

Phenotype: An organism's anatomy, physiology and characteristic behavior; contrasted with genotype.

Phenotype Space: The set of all possible or reasonably possible phenotypes.

Phenotypic Plasticity: All the adaptive phenotypes that an organism with a given genotype can express. This includes behavioral plasticity (modification of behavior) as well as anatomical plasticity (modification of anatomy) and physiological plasticity (modification of physiology) in response to environmental stresses and opportunities.

Pleiotropy: Many features of organisms are constructed of the same proteins and hence produced through the activation of the same genes. Mutations in those genes will therefore affect numerous features all at once, a phenomenon called pleiotropy.

Prokaryote: Bacteria and archaea belong to the biological empire of *Prokaryota*. This category of organism evolved before eukaryotes and gave rise to them. Prokaryotes contain only a single chromosome and have (basically) no internal compartments like a separate nucleus.

Transcription Factor: Certain genes express signaling proteins that bind to other regulatory sequences to stimulate or suppress the expression of other genes. The genes that produce transcription factors can themselves be stimulated or suppressed.

THE NONLINEAR WORLD OF YOSHITSUGO OONO

Most creative thought remains within the familiar, applying well-known viewpoints, assumptions and theoretical frameworks to novel subjects. But a second, less-common form of insight defamiliarizes well-trodden ground so that it appears as if seen for the first time. When Rochefoucauld writes, "We all have the strength to endure the misfortunes of others," or "hypocrisy is the homage vice pays to virtue," he wants us to do a double-take and see things differently. The reader begins the aphorism standing on solid ground and gazing up at the Moon and then, suddenly, finds themself on the Moon looking down.

Eric Smith, the physicist author with Harold Morowitz of *The Origin and Nature of Life on Earth*, and an inveterate user of defamiliarizing language himself, steered me to a true master of the craft, Yoshitsugo Oono.

In the introduction to *The Nonlinear World,* Oono writes,

> The aim of this book is to present a certain "way to think," so inevitably many topics and subjects outside physics must be discussed ... Therefore, this book has comments, remarks and footnotes that may

urge natural scientists to pay some attention to topics of the humanities. The author wishes to count the people on the humanities side among the readers, but free use of elementary mathematics in this book probably requires the writing of another book more suitable for a wider audience.

This essay is my attempt at a non-mathematical translation of this unusual text. Like the first essay in this collection, Oono's work attempts to address complex systems—but from a physicist's rather than a computer scientist's perspective.

THE NONLINEAR WORLD OF YOSHITSUGO OONO

A non-mathematical rendition of
The Nonlinear World: Conceptual Analysis and Phenomenology
by Y. Oono.[79]

Yoshitsugu Oono is a professor of physics at the University of Illinois. Although he has also written a standard college textbook on thermodynamics, his 2013 book *The Nonlinear World* is a literary oddity written in the (nonexistent) genre of mathematical humanism. The main text includes plenty of formulae, but the book's abundant footnotes and side comments reach into far-flung topics, including intuition, value, meaning and the intelligence of dogs and cats. The style is intellectually entertaining, intentionally digressive and a wild ride.

Oono deploys a stream of provocative ideas designed to alter, undermine and reconstitute the viewpoint and thought habits of his audience. In part, he is merely sharing a lifetime of atypical thought, but he also has a pedagogic goal: to prepare his readers to study complex systems from a physics perspective.

Complex systems have proven difficult to define but can be broadly described as systems where the behavior of the whole cannot be directly inferred from the properties of its parts.* Living organisms are the paradigmatic complex systems, and most other common examples are aggregate activities or constructed products of living organisms, such

---

* It is always possible to determine the behavior of a system by operating it, and that may be the only way to predict in any detail what a complex system will do.

as ecosystems, economies, social structures, power grids, ant colonies and the internet. Complexity impinges on what is human because we are complex systems ourselves, and we spend most of our time interacting with other complex systems (i.e., other humans) within the knowledge, thought categories and norms of the complex cultures that shape our complex social structures.

In recent decades, physicists have come to recognize that the young field of complexity science holds mysteries as profound and significant as anything scientists have ever tried to understand. However, progress in the field has been slow. Oono suggests that success might come more easily if researchers better understood and more respectfully regarded the class of scientific theories that can profitably be applied to complex systems.

Oono distinguishes two broad categories of scientific theory: fundamental and phenomenological. Fundamental theories, as he defines them, begin with small units (particles) and elemental concepts (time, space, spacetime fields) and build them up into progressively more substantial compound objects. Theories in this category are typically held out as some of the most profound products of human thinking. Prominent current examples include general relativity, quantum mechanics and quantum field theory. These conceptual frameworks were, in turn, built up from the great fundamental theories of previous centuries, especially Newton's laws and Maxwell's equations.

According to Newton's law of gravity, every particle exerts gravitational force on every other particle, and these forces can be added together, a relationship referred to in mathematics as *linearity*. One can scale up from atoms to grains of sand to planets and stars merely by summing the gravitational effects of individual components; large-scale and small-scale events obey the same formulae. Much the same is true of all other fundamental theories, although sometimes more subtly.* To

---

* It is commonly said that quantum principles apply only to the very small, but it would be more accurate to say that, at larger scales, the more unusual quantum effects cancel out and predictions converge into those of classical physics.

paraphrase Oono, these are microscopic theories that can be inflated to macroscopic size.[80] Fundamental theories lean on first principles.

By contrast, phenomenological theories predict the behavior of large-scale systems without referring to events that occur on much smaller scales Such theories are typically slighted by physicists as "merely" phenomenological, meaning that although they successfully describe phenomena, they do not "explain" them. But this is an inaccurate critique. Newton's theory doesn't explain gravity, and general relativity says nothing about how mass manages to bend space. All theories utilize a floor of primitive units and reach an upper ceiling of explanation, and no theory spans all conditions and circumstances. The primitive units of the game of chess are chess pieces and their rules of movement and capture, not atoms, and the theory of winning at chess owes nothing to Newton, Maxwell or Einstein.

Chess may seem an unfair example because its arbitrary rules can be regarded as a new starting point, a man-made set of first principles. But natural processes also create systems whose primitive units are aggregates and whose behavior is controlled by novel rules that have little to do with the fundamentals of physics. Consider genes. To oversimplify, evolutionary biology studies the schemes that allow genes to "win" the game of natural selection, and while genes are built of chemicals, those chemicals have been so deeply domesticated that they behave more like chess pieces than chemical reactants. There may very well be deep laws of evolution or, more generally, of complex systems, but they will not be rooted in quantum mechanics or the standard model of particle physics.

Purely phenomenological theories can sometimes range as far or further than fundamental ones. Classical thermodynamics—the study of heat, work and temperature and their relation to energy and entropy—is perhaps the most remarkable example. Contrary to popular belief, thermodynamics has never been derived from fundamental laws of physics and it is not clear that it is possible to do so.* The principles of thermodynamics were fully worked out in the 19th century and did not need to be revised *at all* when physics underwent the relativistic and quantum revolutions of the 20th century. Physicists can even assign a temperature to black holes, which must be regarded as a remarkable feat given that all known laws of physics break down within them. While thermodynamics is not a fundamental theory, it is a (nearly) universal one—and perhaps more than universal, given that alternate universes operating under laws different from our own might still follow the laws of thermodynamics. (Thermodynamics may owe more to general informational principles than to physical laws of almost any kind.[81] More on this below.)

Oono highlights the startling universality of thermodynamic analysis in part because he wants physicists to stop glorifying fundamental theories at the expense of phenomenological ones. To understand complex systems, he suggests, we will need to discover the primitive units of complex systems, features that, like temperature, cannot be further subdivided. Oono wants young physicists to understand that the search for regularities in complex systems is just as profound an endeavor as physics' more typical holy grail, the forever-pending "theory of everything" that will unite general relativity and quantum mechanics.

The first four chapters of *The Nonlinear World* establish a general set of ideas for thinking about phenomenological theories. The fifth takes a stab at the phenomenology of complexity, offering fertile ideas that have been influential among some of those who study the origin of life.

---

* It is often assumed that Boltzmann derived the laws of statistical mechanics from Newtonian mechanics, but this is incorrect. On the one hand, he made various non-physical simplifying assumptions so that he could reach the intended result; for example, he assumed, unrealistically, that the motions of bouncing molecules remain uncorrelated. On the other, his derivation remains unchanged under far weaker assumptions than Newton's laws.

Oono has a natively mathematical mind, and most of his examples and illustrations are rendered in formulae. To communicate his intellectual provocations without mathematics, I have been forced to invent stories and find real-world examples that illustrate his points. Despite my best efforts, I have undoubtedly mischaracterized some of the things he is trying to say. I have also intentionally omitted some of his more difficult mathematical points and, no doubt, unintentionally omitted others because they flew over my head. Finally, I arranged the flowing stream of Oono's ideas into what seems to me to be a logical order. This rearrangement was, in part, artifice because the book seems less designed to present a unified thesis than to entrance the reader into a new way of thinking.

Given all these caveats, I nonetheless hope that my rendition at least hints at the scope of Oono's incompletely realized but potentially profound insights.

Let us begin with a question: Why does the universe have laws?

One-Gram Masses on Distant Stars

Discussions of thermodynamics often begin with gas-filled containers whose rigid walls create isolated, closed systems within which gas molecules bounce gaily about. Boxes of gas under high pressure are placed in contact with boxes at lower pressures; hot gases are allowed to mix with cold ones; distinct gases A and B diffuse into one another. In each scenario, random molecular motion causes the combined system to reach an equilibrium state of higher entropy in which pressures and temperatures are equalized and the various constituents are (almost) uniformly mixed.

Inside their rigid containers, gas molecules rebound constantly off the walls and one another. Newton's laws of motion control each bounce, and if a container were truly insulated from outside effects, the past and future positions and velocities of all molecules would be fully determined by their positions and velocities at any instant in time.[82]

But true isolation is impossible: Nothing can block the effect of gravity.

Mathematician Émile Borel is said to have calculated that the motion of a single gram of matter on a distant star will sufficiently alter the complex pinballing of individual molecules here on Earth so that any prediction of their motion beyond one second into the future would be rendered incorrect. The attribution to Borel seems to have been in part apocryphal, but a recent attempt to recreate the supposed calculation concluded that if a single gram of matter four light years away alters its position by a single centimeter, gravitational effects cause predictions of molecular motion on Earth to fail within a mere *microsecond*. Events within the Sun would, of course, cause even more interference.[83] Since the matter composing stars constantly boils and flares, accurate prediction of the future positions of gas molecules is clearly impossible. Practically speaking, molecular motion on Earth is randomized by gravitational noise from space.

In ordinary life, influences far more powerful than those caused by the shifting of distant one-gram masses impinge constantly on everything we observe. This suggests a fundamental question: Why don't chaos and noise drown out everything else? What accounts for the many instances of predictable behavior?

This impossibly deep question has an obvious anthropic-principle answer: If we lived in a universe where nothing could be predicted, we wouldn't have evolved brains. Brains are metabolically expensive and earn their keep only because they make accurate predictions. *Can I successfully cross this river without being washed downstream? Is that object an alligator or a log? If it is an alligator, can it see me?*

The brains of humans and other animals construct explicit representations of the world. But long before they evolved capacities to model the world, organisms already utilized implicit predictive systems. We may ordinarily regard the abstract as more advanced than the concrete, but in one of his lovely insights, Oono points out that "organisms with only primitive sensory organs live in a very abstract world."[84] Single-celled creatures move toward certain chemical stimuli and away from others, but they (presumably) lack the modeling capacity to represent what they are pursuing or fleeing, or even to model the actions of "pursuit" and "flight." They sense and respond to elements of the world without visualizing the world. Evolutionary pressures have tuned such organisms to seek nourishment and respond to threats without giving them the ability to "know" they are doing so.

Animals that possess brains also move toward some stimuli and away from others—but with the aid of interior maps and models. When I seek the source of a tasty aroma, I can visualize a path to the kitchen, avoid obstacles as I walk along that path and reach into a pot to scoop out a spoonful. Presumably, all brain-enabled creatures combine and process the data supplied by their various senses to produce a description of "reality."

Like the evolved responses to sensations that drive animals lacking brains, the world models used by creatures with brains are tuned for evolutionary success. All other things being equal, creatures that rely on bad maps will lack fitness compared to those that build better ones. Over time, natural selection has therefore yielded organisms with increasingly effective models of the world.

Models

A model is a simpler system that captures a subset of features of a more complex one and does so in a useful way. Newtonian gravity models falling objects as if they were moving in an absolute vacuum. This simplifying assumption is universally false, but the Newtonian model based on it makes excellent predictions in many circumstances of interest; it can do so because the interfering effects of friction are small enough that they can be ignored.

There are (typically) many possible ways to model the same phenomena. When ancient Greek astronomers attempted to represent the motions of planets in the sky, unquestioned philosophical assumptions compelled them to think in terms of perfectly circular, constant velocity orbits around a central, unmoving Earth. To extract accurate predictions from this geocentric model, Alexandrian astronomer Ptolemy of the second century AD was forced to add smaller circles (epicycles) to the larger circles of overall orbits. At the limit of the observational precision permitted by the astronomical technology of the time, this "Ptolemaic system" provided reliable predictions of planetary positions. It was an accurate model, if not a simple or elegant one.

More than a millennium later, Copernicus proposed an exceedingly elegant heliocentric model in which planets revolved around the Sun. However, like the ancient Greeks, he problematically assumed that nature would inevitably rely on perfect circles and uniform motion. This elegant system made bad predictions, and Copernicus, too, was forced to add epicycles; but even with their aid, his approach never yielded results as accurate as the Ptolemaic model.

Kepler picked up on Copernicus's insight and, finally abandoning reliance on circular, unvarying motion, successfully modeled the motion of planets as elliptical orbits in which the velocities of each planet systematically varied. The Keplerian system supplied much more accurate predictions than any of the others, but it wasn't particularly elegant. Newton overcame this esthetic failure when he re-derived Kepler's description from a few mathematically simple and intuitively plausible laws of motion and gravity.

Effective, if not necessarily elegant, world-modeling is built into our cognitive and sensory functions. For example, as extensive perception research has documented, humans perceive little in the way of raw sensation. The incoming sensory stream is immediately analyzed into features we have evolved to find important. Eyes see edges, rapid motion and salient objects, not raw video. Rather than experiencing the world directly, we interact with our evolutionarily derived models of reality.

When I walk my dog, I pass through a kind of cartoon populated with "houses," "streets," "cars" and "pedestrians" rather than comprehensive visual data. My dog's visual model of the world is likely quite similar, though filled with rather more fire hydrants, stop sign poles and ambient squirrels than parked cars and houses.*

Both Kepler's analysis of planetary motion and the typical visual experiences of dogs and humans utilize what can be roughly described as a noun-verb model of reality. This perspective analyzes the world as if it contains discrete "objects" that move through space, whose positions can be tracked and predicted and that possess edges and characteristic properties.

But this model is full of assumptions that often fail on close inspection. Are water droplets object-like? Unlike pebbles, they merge and separate. Given that the roots of a mature tree are typically penetrated by fungi that may extend for miles, where does the tree end, exactly? When a cow reproduces, the organisms in the cow's stomach reproduce, too. Are those organisms part of the cow-object? Does the boundary of a living human include its non-living fingernails? What about clothes? (In fiction, invisibility rays usually—but not always—work on clothes, too.) Where does a mountain end and its foothills begin? How many cars are there in a junkyard that includes partial, broken, non-functional cars?

Even this brief attempt shows that it is exceedingly difficult to define what we mean by "an object." But a precise definition is unnecessary because we all know what we are talking about. Evolution has found it useful to construct brains that model the world as if it were composed of object-like constituents, and we all take that object-ness for granted.

---

* She also moves within an olfactory world model that lies beyond my comprehension. Oono writes, "When we compare the natural intelligence of dogs and that of human beings, we tend to believe there is a tremendous difference. However, if we take into account the very high processing ability of olfactory information by dogs, it is more reasonable to think that the difference in the natural intelligences should be understood as difference in directions but not in magnitudes. ... To the zeroth order the intelligence of all the mammals should be approximately 'the same.'" *The Nonlinear World,* p. 276.

Evolution has additionally provided us with cognitive structures that group objects into sets whose members possess similar properties. Rocks are hard, predators bite, bananas are tasty and birds move rapidly out of reach. No two hyenas are identical, but they have enough in common that classifying each individual hyena as an "example" of the general "type" "hyena" offers obvious advantages in world-prediction.

Half a billion years ago, fish hunted other fish, and it would seem likely that the brains of hunters and hunted converted sensory perceptions into representations of objects and categorized them as threat and prey. Given the evolutionary depth of this style of world modeling, it should be no surprise that when physicists studied the extremely small and found that object-ness fails in the quantum world, they experienced intuitive confusion.

Consider the history of the concept of "the electron."[85]

By the late 19th century, physicists had become aware that something rather like a small particle carried electrical charge. Some physicists believed that charges were carried by electrically charged chemicals known as ions; thus, electrons were first referred to as "electrions." Other physicists proposed that charge was carried by small, localized waves that obeyed Maxwell's equations of electromagnetism, while yet others hypothesized small rotating elements locked into position in the ether.

The notion that electric charge was carried by a very small but otherwise ordinary object coalesced in 1897 when J.J. Thomson created an electron beam and showed that it could be understood as being constructed of discrete elements. In 1909, Robert Millikan measured the charge of a single "electron" and demonstrated that all electrical charges appeared in multiples of this fundamental unit. The first phase in the intellectual history of the concept of the "electron" had been completed.

Electrons continued to enjoy their object-like existence for the next 25 years. Like other objects, they were assumed to possess mass, location and velocity. Their only deficiency with regard to typical objecthood lay in the matter of spatial extension: electrons seemed to be points. But perhaps they were merely *very, very small* objects.

And then came quantum mechanics.

It had been known for some time that light displayed both particle- and wave-like characteristics. In 1924, Louis de Broglie generalized this well-known (if already problematic) understanding into the mind-boggling idea that *all* small particles are simultaneously waves and particles. Using this idea of "wave-particle duality," he proposed that electron orbits around atomic nuclei are exact integer multiples of electron wavelengths, thereby producing standing waves. This suggestion worked out surprisingly well. Predictions based on wave-particle duality closely matched experimental results.

The idea that an object can be both a wave and a particle is weird enough, but things got even weirder when Schrödinger announced his eponymous wave equation in 1926 and Heisenberg his uncertainty principle in 1927. Electrons, scientists now understood, lack two characteristics that anything described as an object must possess at any given time: a precise location and velocity. Instead, the phenomenon referred to as "an electron" spreads out in a probabilistic cloud. As an even weirder detail, these probabilities emerge as imaginary numbers.

The predictions made by these two mathematically equivalent approaches were abundantly verified, and there is no doubt today that we live in a quantum rather than a classical world. But quantum mechanics continues to cause a great deal of consternation. It makes no intuitive sense that certain "objects" lack simultaneously precise positions and velocities and that, instead, they possess nothing more than a probabilistic range of positions and velocities. The idea is bizarre. It feels paradoxical.

But there is no paradox.

Humans and other animals possess cognitive structures that model the world into objects and their actions. While this is an excellent model in normal life, it fails at sufficiently small scales. If evolutionary pressures had required us to model electrons, the ideas of quantum mechanics would feel perfectly intuitive. We weren't so required, and so they don't.

Phenomenology and Understanding the World

When a gibbon leaps from branch to branch, counting on the dip and rebound of each temporary handhold to power its dizzying airborne glide, it displays considerable skill at solving the differential equations that describe motion under the effect of gravity (along with others that predict the elastic behavior of branches). Gibbons *understand* gravity. And yet, they know nothing about the differential equations their brains are using analog methods to solve. In Oono's terminology, gibbons apply an effective phenomenological version of Newtonian mechanics within the range of events that matter for their lives. Referring to what seem to be his two favorite animal species, he writes:

> To understand the world phenomenologically is to recognize such general mathematical structures behind various phenomena. However, no mathematical knowledge is required to recognize them. Even dogs and cats understand that classical mechanics is governed by a second-order differential equation, so they rarely fail to eat or are rarely eaten. Of course, to express consciously what one knows (or what is embodied) is difficult, so there is no differential equation in the dog's brain. However, most human beings are not different from dogs in this respect.[86]

Unlike the understanding of the phenomenology of motion built into our bodies, Newtonian gravity expressed in mathematics is a fundamental theory that can describe the orbit of the Moon as competently as the behavior of falling stones. Such theories have pride of place in science, especially in physics, where they are regarded as among the greatest pinnacles of human intellectual accomplishment. Fair enough; human beings alone have developed fundamental theories. However, in real life, fundamental theories are frequently too difficult to apply, and working scientists have found they must settle for phenomenology. All of chemistry can, *in principle,* be deduced from quantum mechanical principles, but the mathematics of doing so remains unsolved, and the science of chemistry uses formulae of convenience that incorporate parameters determined.

Nonetheless, that "in principle" provides chemistry with a direct connection to a fundamental theory. There is no obvious equivalent in complex systems. Consider machine-learning programs, systems that are becoming increasingly important to our lives. At the time of this writing, large language models such as ChatGPT are some of the most well-known examples. By "training" on much of the internet, ChatGPT acquired an ability to predict what words are most likely to follow other words as well as to classify writing samples into styles, genres and other categories. The relatively straightforward algorithms by which ChatGPT constructs its operation rules unexpectedly permitted it to carry on conversations with persuasively human-like fluency and answer questions that it had previously been assumed would require human-like intelligence. Researchers did not know that these capacities would emerge until they observed them and it is precisely such emergence that suggests large language models should be regarded as effective models of a complex system: the human brain (or, at least, of language, which itself models aspects of human brain function). But neither ChatGPT nor the complex system it partially models owe anything to quantum mechanics.

The exploits of machine-learning systems are difficult to predict because the models they create include vast arrays of numbers that are opaque to human understanding. Some researchers have suggested that this opacity will cause the operation of machine-learning systems to remain forever incomprehensible. But I disagree to this extent: I predict that in the not-very-distant future, machine-learning researchers will *talk* as if they understand the systems they build.

It seems plausible that no amount of knowledge of microscopic states—the content of machine registers and the like—will contribute significantly to predicting the macroscopic behavior (outputs) of large language models. However, this implies only that researchers will fail to discover a kind of fundamental theory that predicts the behavior of these models. The history of our relationship with other minds suggests that researchers will invent phenomenological theories that successfully predict some of the behaviors of large language models and other machine-learning systems. And once those theories become sufficiently familiar, researchers will begin to use the words "I understand how that system works" when they speak of their creations.

I predict this because we use the language of understanding when we speak of other minds.

Theories of Mind

Many social animals can simulate the mental states of other members of their species and thereby predict their behavior. It behooves a non-dominant chimpanzee interacting with an alpha to analyze the state of knowledge and the likely behavior of the alpha so that it can avoid punishment. Is the dominant able to see that stash of fruit? Can I grab some without drawing the alpha's attention? If not now, will I be able to snatch it when he is distracted by that boisterous juvenile approaching from behind a bush?

The ability to model mental content is called a "theory of mind," and chimpanzees are pretty good at it. Like us, they are equipped with an advanced supercomputer installed between the ears that uses neural networks and other systems to engage with the world. For social animals, interactions between members of the same species feature highly in the specifications of those supercomputers.

However, it isn't realistic to expect that one chimpanzee's theory of mind will correspond to the detailed neurological structure and function of another chimp's brain. But it doesn't have to. To be useful, a theory of mind need only predict certain behaviors of interest more accurately than chance.

Humans also model one another's minds. Philosophers call our method of doing so "folk psychology," a kind of stick-figure, commonsense analysis of the behavior of our fellows (and of ourselves). We predict human behavior by assigning such characteristics as motivations, desires, temperament, beliefs, tendencies and habits, and then calculating how they interact. I know that I accidentally *insulted* my coworker in public; he is now *angry* at me, and unless I engage in actions to remediate my error, such as *apologizing* and *praising* him to his *superior*, he may find a way to *get even*. Police detectives expect most murderers to have *motives* for their crimes.

Again, it would be surprising if folk psychology closely matched actual brain states. The folk psychology model suffices because it often (although certainly not always) makes reasonably accurate predictions—it is a successful phenomenological model. But we do not feel it as a mere model. When we think about people in folk psychology terms, we feel that we *understand* them. We have even come to believe that we directly sense such things as beliefs, desires and intentions in *ourselves*, as if we were looking into the operation of a watch. But it seems unlikely that there exist neurophysiologic sliders that control the intensities of desires, registers that hold beliefs or global variables that set temperament.* A deep structure of that kind isn't consistent with the much more holistic fashion in which brains are believed to work.

Given the success of folk psychology at predicting human behavior, it seems plausible that future computer scientists will invent

---

* This is not to say that folk psychology *always* fails to correspond with actual physical states. When I accidentally bump my dog, she looks at my face to see if I am angry because if I spanked her in anger, she knows she should respond differently than if I only bumped her by mistake. "Anger" is an example of a folk psychology diagnosis that might indicate an actual physical state, in this case, a certain kind of stereotypical emotional arousal, perhaps hormonal in nature, common to many social mammals.

comparable "folk psychologies" of large language models and that repeated successful uses of such models will inspire machine-learning researchers of the future to feel that they *understand* how such systems work. The difference between predicting and understanding is largely esthetic.

Isolated Systems

An "object" is an element or collection of elements that is partially isolated from its surroundings. Earth as a whole exchanges gravitational influences with Jupiter, but very little that occurs on the Earth would be changed if Jupiter suddenly disappeared. In general, to model any subset of reality, that subset must be somewhat isolated from external influences. If gravity didn't fade rapidly with distance and faraway stars could influence the motions of cannonballs, Newton's law of gravity would never have been discovered. Similarly, thermodynamic relationships will not hold when sufficiently large amounts of energy are being added to or subtracted from a system. The need to isolate manifests in the explicit design of scientific experiments, in which extraneous influences are limited while the behavior under study is probed, prodded and measured.

More precisely, experimental design seeks to hold extraneous influences *constant*. A system need not be totally isolated for its behavior to be susceptible to modeling, but external influences that do make an impact must appear as constant parameters or at least as slowly changing, predictable variables. The early Earth was frequently bombarded by substantial objects from space, making it a seething hellscape; weather prediction would have been somewhat more of a challenge than it is today. But modern Earth is largely left alone. True, our weather is powered by the distant Sun, but the Sun's output is sufficiently stable, and the variations in solar energy entering the atmosphere in specific regions due to Earth's rotation about its tilted axis and its orbit about the Sun are sufficiently predictable that those factors can be worked into weather models as simple parameters.

Physical separation is a great aid to system isolation, and it occurs commonly in our universe because attractive forces cause materials to clump. We live on one such clump. The dim light of the fixed stars provides a frame of reference for navigators, but beyond that informational resource, the main interaction between our planet and the remainder of the universe occurs in the form of reasonably stable solar radiation. Lunar gravitation adds an additional but much less important external influence by producing tides.

In addition to mere distance, Earth is additionally isolated by *natural barriers*. Jupiter's intense gravity is thought to have cleaned up most of the early solar system's small objects, thereby reducing the frequency of bombardments from space. Earth's magnetic field largely blocks high-energy particles emitted by the Sun, distant supernovae and other sources. Atmospheric ozone captures most solar ultraviolet radiation, and Earth's entire atmosphere burns up space rocks smaller than about 25 meters in diameter. These barriers enhance predictability down on Earth; if cosmic rays and high-frequency ultraviolet radiation reached Earth unimpeded, the DNA of parent organisms would much less reliably contribute to the information resources of their offspring.

Similarly, while Earth's interior is dangerously hot, its crust acts as an excellent insulator, and it is only through events such as volcanic eruptions and earthquakes that Earth's internal heat significantly impinges on surface life. Of more immediate effect are the disturbances in the Earth's atmosphere known as weather. We find rain and high winds problematic and isolate ourselves from them (and other intrusions) by constructing *artificial barriers* such as houses. Beavers do much the same when they construct their lodges. There are fewer variables to contend with inside a house or a beaver lodge than outside.

At a more primeval level, and long before the evolution of beaver lodges, early living organisms evolved the ability to isolate themselves into partially closed systems by means of such useful contrivances as skin, cell walls and cell membranes. Within those protective and confining barriers, living organisms employ homeostatic mechanisms to further damp down the effects of perturbations from outside. Some multicellular animals maintain a fixed internal temperature regardless of the outside temperature (within limits)—a form of climate control.

The use of barriers is essential to living things because life occupies a crowded realm where nothing is far from anything else. Physical distance separates Earth from the nearest star but does not protect bacteria from constant interaction with other bacteria. Organisms must manage in a jam-packed world.

The Realm of the Densely Packed

It's often said that the universe is largely empty, but this commonplace is false at the scales relevant to biological organisms.

At the energy levels of ordinary chemistry, atoms behave like spherical objects, albeit with soft edges, like dandelion puffs. Equal-sized spheres shoved against one another fill 74 percent of the available volume, leaving 26 percent empty in the interstices. The actual "packing fraction" of atoms or molecules in solids may slightly exceed 74 percent because electron-sharing can reduce atomic diameter; at the other extreme, some solids pack their molecules less tightly, but seldom with a packing fraction lower than 30 percent. The packing fraction of most liquids also falls somewhere near 30 percent.

To visualize how dense this is, imagine an Olympic-sized swimming pool 50 meters long, 25 meters wide and 2 meters deep, constituting a volume of 2.5 million liters. Assuming the average volume of a person is about 65 liters, if the pool is packed with people at a packing fraction of 30 percent, 12,000 people would fit in a single pool.

The inside of a cell is similarly crammed, but with molecules. Water composes a kind of "ether" for life processes; between water molecules, there lie innumerable small organic chemicals, medium-sized cofactors, long strings of amino acids (proteins) and gigantic strings of nucleic acids (DNA). Cell interiors are wall-to-wall chemicals, and this crowdedness continues when scaling up: About as many cells are packed into a human body as there are atoms in a cell. Outside of living organisms, inorganic molecules clump into dense rocks, mountains and continents, and continents float on deeper Earth layers. A realm spanning 17 orders of magnitude from the diameter of an atom to the diameter of Earth is crowded on every scale.

The packing fracture of molecules in Earth's atmosphere is much lower, about 0.1 percent at standard temperature and pressure, but below the size of atoms and above the size of Earth, space begins to be truly empty.

The total number of protons that could fit into a hydrogen atom is on the order of 10^{15}. Since there is only one proton in a hydrogen atom, the packing fraction in terms of protons is 0.74×10^{-15}. This is about the equivalent of one person in the volume of water contained in 25 billion Olympic-sized pools, and one trillion times less packed than the air we breathe. Protons do not often interact under earthly conditions because they are too far apart.

The emptiness of the gap bracketing the densely packed world from above is nearly as extreme. A sphere with a radius equal to the average orbital distance of Neptune encloses about 10^{30} cubic kilometers. Since the volume of the Sun is about 10^{18} cubic km, the total number of Sun-sized stars that could fit into a solar system-sized sphere is on the order of 10^{12}, a number that exceeds the number of stars in the Milky Way galaxy.[87] But since there is only one star in the solar system, our local piece of the universe has a packing fraction on the order of 10^{-12}, or one person in the water of 25 million Olympic-sized pools.

The dense realm of life is bracketed by empty space. Within the dense world, clumping and barriers commonly produce isolation. And when these fail, systems are often effectively isolated by scale.

Scale Interference

> *"For thousands more years the mighty ships tore across the empty wastes of space and finally dived screaming on to the first planet they came across—which happened to be Earth—where due to a terrible miscalculation of scale the entire battle fleet was accidentally swallowed by a small dog."*
>
> Douglas Adams, The Hitchhiker's Guide to the Galaxy.

In a fight between a professional heavyweight boxer and a flyweight, the lighter contender will always lose, even if they are the greatest in the world in their weight division. If he were given the freedom to punch away unmolested for as long as he wanted, a flyweight boxer could knock out Tyson Fury, but the weight ratio between the heaviest and the lightest fighting classes is merely 2:1. In the current discussion, scales are estimated in magnitudes, and 10:1 is the smallest difference considered. Not even the burliest toddler could concuss Fury if given a year to try; make the difference two orders of magnitude, and it becomes a battle of Fury vs. a teacup Chihuahua. Physical battles across scale differences of an order of magnitude are laughable because the small creature simply can't affect the larger.

Sometimes, however, systems operating at much smaller scales can metaphorically punch above their weight. The venom of a three-pound inland taipan can fell any human, and while the bite of a 0.035-ounce black widow spider couldn't kill Fury, it would disable him for days. We find such disproportionate interference from a smaller scale disturbing; while we may fear lions, tigers and bears, we *shudder* at snakes and spiders.

And at least we can see spiders. Sources of death too small to see disturb us even more; hence our often irrational responses when threatened by plagues. Radiation and chemical and biological weapons horrify us similarly; the 1925 Geneva Protocol bans gas attacks but not bullets. The most extreme of disturbing invasions from one scale to another is the objectively nonexistent one of ghosts, creatures said to be non-corporeal. Tigers frighten; spiders horrify; plagues arouse superstitious fears; and the literal superstition of ghosts can cause shrieking terror.

Interference from scales far greater than our own produces different but nonetheless powerful emotions. We respond with awe to lightning, earthquakes, tornados, floods and hurricanes and call them acts of God.

The typical isolation of a given scale from much smaller ones depends on the common, but not universal, phenomenon of *linearity*.

Linear vs. Nonlinear Effects

Linear relationships are proportional—twice the input leads to twice the output. Ten times as many people need 10 times as much food. Walking 10 times as far takes 10 times as long. During a steady rain, twice as much water collects after an hour than after half an hour. Because many useful elements of our everyday existence operate according to linear rules—at least approximately—we humans (and animals) understand linear relationships more fluently than any other type of mathematical dependency.

Linearity limits the effect of the small upon the great. If the maximum force Fury's fist can impart by repeatedly punching an aircraft carrier is several orders of magnitude less than the other forces that impinge on the carrier, such as the impact of waves, all his efforts will produce no more than a minor vibration.

Linearity also limits prediction error. It is easy to predict the behavior of linear systems to whatever accuracy one desires because a 1 percent error in measurement at the beginning of a linear process leads to a 1 percent error in the predicted outcome. But, as this essay's title reminds us, our world is far from uniformly linear, and nonlinearity can permit microscopic events to intrude into daily reality, making prediction difficult or impossible.

If there are X cases of infection on day zero, and the number of infections doubles every five days, then a 1 percent error in measurement on day zero leads to a 2 percent error on day five, an error greater than 100 percent after a month, and, if exponential growth continues—although it can't continue indefinitely—a 26,000 percent error in three months. Increased accuracy in estimating the subsequent behavior of such exponential systems requires exponentially more precise data about the starting conditions. This phenomenon is called "sensitive dependence on initial conditions," and it explains in part why no COVID-19 models were as accurate as we would like models to be. It would require unreasonable precision in the knowledge of the initial conditions of exponential processes to make reasonably accurate predictions of their future course.

As a rational response to this difficulty, humans have evolved responses to exponential processes that conflate any number above zero with infinity. This simplification manifests in the psychological phenomenon of "contamination." If I place a drop of rotten meat juice in a glass of water, stir the liquid and remove one drop and add it to another glass of water, and then repeat the process twice more, an automatic feeling of disgust will make it difficult for me to drink from the glass. Three such dilutions decrease the number of pathogens by a factor of perhaps 10 billion, far more than enough dilution to allow safe drinking. Nonetheless, even when contaminated water is diluted this much, the idea of drinking it remains revolting.

This intuitive response is perfectly rational because pathogens grow like the doubling grains of rice on the chessboard in the famous parable, and even a single invisible bacterium can cause death if allowed to multiply. Since we can neither count pathogens without a microscope nor intuitively estimate exponential growth, it is perhaps best to avoid contact with any pathogens at all. Of course, this contamination heuristic can fail, just as it did during the pandemic when (as immortalized by Kate McKinnon in a famous *Saturday Night Live* sketch) people bleached the outsides of bags of Doritos two or three times before permitting them into their homes.[88]

Exponential processes are a common means by which the small can interfere with the great. But exponential growth still preserves ordering relationships. If exposure to seven virions is enough to cause severe illness in a high percentage of patients, exposure to 14 virions is even more likely to do so, up to the point of "clipping," in which the viral load is sufficient to cause severe illness in everyone. This is the argument for wearing masks and using Plexiglas barriers to reduce the spread of COVID-19. Many people still find the value of such precautions non-intuitive because their intuition tells them that *any* exposure is the same as maximal exposure. But that is incorrect: The risk of COVID-19 transmission is sensitive to dosage.

However, not every form of interference from smaller scales preserves order. Non-ordered, non-linear relationships occur in chaotic systems where small changes in initial conditions lead to wildly fluctuating results. The world contains many such systems, and their existence makes it surprising that we can predict what will happen in the world at all.

Chaotic Systems

Imagine a pinball machine played only by pulling back on the release lever, without use of flippers. Extremely tiny changes in the velocity at which the pinball is released will result in widely varying scores. Pulling the lever farther back doesn't bring about consistently higher or lower scores; rather, greatly different scores, both higher and lower, may occur in response to minimal changes in the strength of the initial spring-loaded impact. This is non-ordered nonlinearity; more formally, nearby trajectories both diverge and cross over. Pinball is a chaotic system.

When changes far beneath the scale of observation can produce large changes in observable outcomes, even approximate prediction may be impossible.

The behavior of molecules in a gas resembles a pinball machine, which is why the movement of one-gram masses on nearby stars makes it impossible to accurately predict their individual motions. The timing and path of a lightning bolt; the path of a rivulet of water on a smooth windshield; the intervals between earthquakes; and, over longer timescales, the motions of the planets in the solar system (which cannot be predicted past some number of millions of years) also illustrate chaotic, unpredictable behavior. While we can state confidently that a lightning bolt will eventually strike and that the solar system is likely to fly apart in time, we can't predict when.

Note that pinball machines and the other examples just given obey all the normal laws of physics; their behaviors are, therefore, deterministic in an ultimate sense (whatever that means). However, because microscopically small changes in initial conditions dramatically affect the future behavior of chaotic systems, extraordinarily fine control of input force would be required to produce any desired outcome. When hands are used to launch a pinball, force generation by muscle cells varies continuously and can't be stabilized beyond a certain, inadequate point. In every practical sense, the outcome is unpredictable.

Sensitive dependency on inputs is the basis of the famous butterfly effect. The term originated in the science-fiction short story by Ray Bradbury, "A Sound of Thunder," in which a time traveler accidentally kills a butterfly during the era of dinosaurs, setting off a series of changes to the timeline that reach forward 100 million years and alter the course of an election in the time traveler's own day. Based on this idea, it has been suggested that the flap of a butterfly's wings in Brazil might influence the location of a tornado in Texas.[89]

This claim is unlikely to be realistic as stated. Not all changes in initial conditions lead to exponentially diverging outcomes because one or another analog of friction may act to dampen their effects. Those who wish to improve the world have learned to their sorrow that it is difficult to do so; small acts of kindness reliably fail to transform societies unless a great many people simultaneously engage in them. Most of the acts a time traveler might commit would plausibly fail to produce any persistent effect because history is insensitive to small changes. Nonetheless, it isn't difficult to envision scenarios in which small random acts have large effects, such as if an errant driver were to run over a young Adolf Hitler.

Both exponential and chaotic processes are forms of nonlinearity. Exponential relationships magnify small differences in initial conditions, while chaotic relationships both magnify and shuffle. Both processes cause infinitesimally different starting trajectories to radically diverge and permit events at relatively small scales to affect events at larger ones. Mutations in living things illustrate this divergence: Small molecular changes to DNA can cause enormous effects on large organisms, including preventing the formation of any organism at all.

To put it another way, linear systems may turn up the volume of noise, but nonlinear systems can utilize noisy inputs to bring something entirely new into the world.

The Almighty Noise

> *"In contrast to the usage of this word as in 'what he says is a mere noise,' there is no connotation … that what we call noise is unimportant or trivial. Rather, noise is regarded (almost with awe) as an information source beyond our reach."*
> Y. Oono.[90]

Noise consists of unpredictable perturbations impinging on a system of interest, usually arising from levels smaller than the observation scale of that system. There is nothing truly random about noise; it carries information from other scales and from far away. However, if that information is uncorrelated to the system of interest and combines multiple distant sources, the net result satisfies most operational definitions of randomness.

The effects of noise on linear systems are often manageable because small perturbations produce small effects. But in systems with sensitive dependence on initial conditions, noise may transform the system's behavior, functionally introducing new information into the world. This effect can be seen in the common phenomenon of *symmetry breaking*.

In physics, "symmetry" refers to physical or mathematical features of a system that are preserved under a defined transformation. A container full of liquid water at equilibrium possesses (very nearly) constant density at all points within that container. This feature demonstrates "translational" symmetry, the persistence of a physical feature despite movement in space. When a liquid condenses into a solid, it loses translational symmetry: Frozen water is lumpy.* Such loss of symmetry is called symmetry breaking.

To see how noise introduces information when symmetry is broken, imagine a perfectly round steel ball balanced at the top of a roughly conical hill. Once placed in that (unstable) position, the ball is equally likely to roll down in any direction; the future of that system is symmetric with regard to direction. However, the ball can only roll down a single path, and once it has gone down, it can't get back up to try again: directional symmetry has vanished. Because there is no intrinsic superiority of any single direction, it takes only an infinitesimally slight difference in input to "choose" which way the ball rolls. Due to the extreme sensitivity of such a prepared system, the "choice" is made by noise. The direction "chosen" is new information added to the world. The Almighty Noise has spoken.

This example is artificial and a bit absurd, but symmetry breaking of a mathematically similar kind occurs frequently in nature, often in conjunction with phase transitions. One classic example is the tendency of ferromagnetic materials to spontaneously magnetize when they are

---

* What if the water freezes into a perfect crystal? Translational symmetry is still broken, but this time more subtly. In water, constant small-scale molecular movement tends to make each portion of resting water identical to all others, at least over any but very short time intervals. In solids, there are fixed "holes" between atoms where density is zero, alternating with atoms of non-zero density.

cooled. It is easy to predict the strength of the ferromagnet in advance, but its ultimate north-south orientation remains as unpredictable as the direction a perfectly smooth ball will take when it rolls off a symmetrical peak.

In a cooling system, many forms of symmetry breaking occur in succession. Shortly after the Big Bang, strong, weak and electromagnetic forces are thought to have existed as a single force. But as the universe cooled, they separated into different forces.[91] These were symmetry-breaking events that produced arbitrary values in the ratios between force strengths and behaviors. Initial conditions only minutely different from those that produced our universe would have resulted in different outcomes. Parameters that come into being during symmetry-breaking events can't be deduced from the fundamental theory and must be added in as observed measurements; they are contingent and historical products of the Almighty Noise.

Symmetry breaking did not cease with the formation of the basic forces and constituents of the universe. The entire evolution of life can be viewed as a series of symmetry-breaking events in which (to oversimplify) mutations that could have gone one way went another and never turned back.

The existence of symmetry breaking limits the ability of fundamental theories to explain the universe. Even if we had a full-fledged theory that united quantum mechanics and relativity and that also had access to infinite calculating power, we still couldn't predict the outcome of symmetry-breaking events. This is true on a practical basis because systems with sensitive dependence on conditions require impossibly precise measurements of those conditions to predict outcomes. It might also be fundamentally true because the Heisenberg uncertainty principle limits our ability to obtain desired detail regarding the states of small objects. A perfect theory of the forces and constituents of the universe might not tell us much about what the universe looks like because infinitesimal variations lead to radically different histories.

However, despite the profound unpredictability at the heart of reality, we can predict many things quite well. One of the most effective ways our minds tame unpredictability takes advantage of the power of averaging. Averaged data may behave in predictable ways even when the underlying processes producing that data are thoroughly unpredictable.

Averaging

Even though moving single-gram masses on Proxima Centauri, our Sun's nearest neighbor, make it impossible to predict the behavior of *individual* gas molecules, we can acquire highly useful information by estimating the average of the squared velocity of each molecule multiplied by its mass. This converts numerous chaotically changing variables into a single, slowly changing macroscopic variable that is both important and (relatively) easy to measure. The colloquial name for this emergent parameter is *temperature*.

When two objects at different temperatures are placed in contact with one another and allowed to remain in contact, the laws of thermodynamics tell us that they will gradually adopt a new temperature between the initial temperatures of each. This remains true whether the objects are composed of solid, liquid, gas, plasma or the degenerate matter of neutron stars, so small as to be influenced by quantum mechanical phenomena or sufficiently massive as to bend space into a black hole.[92]

Among its many effects, temperature influences the rate of chemical reactions. Reactions that proceed at reasonable speeds at medium temperatures will slow to a crawl when cooled and accelerate explosively at higher temperatures. Since living things are largely chemical machines, temperature matters a lot, and it is no accident that organisms evolved mechanisms to estimate it. The scientists who invented thermometers piggybacked off evolution-derived sense organs. They might not even have been able to pull it off if they couldn't use temperature sensation as their starting point.[93]

Standing Back

Thermodynamics provides a somewhat brute-force example of how lawfulness can emerge from chaos. Human (and animal) minds also use more subtle techniques to identify patterns in the noisy universe. One of the most common and useful is *standing back*, or *coarse-graining*.[94]

This technique involves systematically reducing the resolution of a dataset to reveal large-scale patterns. When a blown-up photo is studied from inches away, it appears noisy and indecipherable, but the intended image pops out when one steps back a few feet. The remaining detail is coarse-grained rather than fine-grained; it is pixelated, smoothed out and approximated, but systematic data loss often makes large-scale patterns easier to see.

History differs from journalism because journalists writing about current events stand too near their subjects to recognize trends. When studied up close, human events exhibit sensitive dependence on factors such as personalities, accidents, famines and years of excess or insufficient rainfall. But when they are examined from a distance, the talents of individuals and random decades of drought smooth into large-scale patterns. If Sargon of Akkad hadn't built the first Middle Eastern empire, someone else would have done so, and it is not obvious that subsequent human history would have been that much different. Events nearer to us in time are obscured by noise and affected by personalities. Napoleon and Hitler seem to have made person-specific impacts on history—and so they did—but millennia from now, one can be sure that their actions will have been coarse-grained and curve-fitted into long-term trends.

Coarse-graining is a prerequisite for classifying objects into categories. No two horses are identical, but when observed from afar, they are identical enough. An antelope perceives the edges of a moving silhouette at twilight and assigns it to the same class of objects as the rippling, spotted bodies of leopards observed at noon. Biologists of Darwin's time did much the same when they referred to "species," an idea that is easy to work with but devilishly difficult to define.[95]

Like averaging molecular velocities to yield temperature, standing back surrenders small-scale detail in favor of large-scale patterns. Humans and other animals are compelled to use such methods because the universe has too much going on; if we tracked all the details, we would drown in data. Fortunately, it is usually possible to ignore a great deal of data and draw reliable conclusions from a selected subset.

But there exist certain extraordinary states of matter where (almost) every detail matters.

Special Preparations of Matter

Imagine that a cup falls off a table, hits the floor, bounces and shatters.

Like many other common events, this transition from order to disorder is irreversible. Shattered cups never leap from the floor of their own accord to reassemble themselves intact once more on a table. Even if we wanted to engineer such an event, we couldn't. Events move from past to future along a one-way street.

But the apparent irreversibility of events is an artifact of coarse-graining, our necessary inattention to fine detail. If we could track the position and velocity of every molecule involved and play a reversed movie from the moment of the shattering, we would observe the ordinary laws of physics causing a jumble of flying fragments to strike the ground, reassemble on the bounce, and fly upwards with just enough kinetic energy to gently set a pristine cup back on the table. If we were to run the movie starting 10 minutes after the shattering, we would watch air molecules from across the room converge on the fragments in a bizarrely coordinated dance—one that is affected by moving masses on Proxima Centauri—and lift ceramic shards off the ground in just the right way so that when they fall after being lifted, they will coalesce into a cup.

Reversed movies of crashed cups look silly, but the laws of physics are fully reversible,[96] and if we could follow such movies at a microscopic level, we would observe that each element of matter merely follows Newton's laws. This seemingly impossible net effect can only emerge if atoms and molecules are pre-placed in just the right way. If we wanted to demonstrate the un-shattering of a cup moving forward in time, we would have to measure the exact positions, velocities and internal energy states of every molecule in the room (and some on Proxima Centauri, too) and then set up a scenario in which all these positions are duplicated, but with reversed velocities.[97] However, errors in perhaps the tenth or twentieth decimal point would suffice to sabotage the demonstration. Time reversibility is infinitely fragile. Within the range of possible positions and velocities of all relevant molecules, only a vanishingly small set of arrangements would work.*

Nonetheless, the exceedingly special, highly tuned, delicately arranged circumstances sufficient to reconstitute shattered cups, reverse explosions and reanimate life if time were to be reversed must surround us, given that we constantly observe the opposite happening in our forward-time universe.

The time-reversibility of physics leads to Loschmidt's paradox, a famous insight that confounds any simple understanding of the arrow of entropy. The statistical arguments used to indicate that entropy increases going forward in time prove that entropy should also increase going backward in time; therefore, both the past and the future should possess greater entropy than the present—a paradoxical result. This is a deep subject that has caused considerable philosophical headaches and for which no persuasive solution has been found.[98] But one thing that can be said is that entropy would remain constant if every detail of a physical system were tracked. Entropy is only perceived to be increasing in the coarse-grained world; like all thermodynamic phenomena, it is an artifact of deliberately elided detail.

We can't tune matter with sufficient precision to un-crash cups, but we handily utilize a different form of specially prepared matter that also

---

* It's easy to show that there exists more than one set of molecular positions and velocities that could lead to the same macroscopic event: Simply tip cups off the same table more than once.

includes carefully arranged, highly unusual small-scale configurations carefully tuned to yield macroscopically apparent outcomes. Unlike the delicately reversible states that follow shattered cups and mixed gases, these preparations are highly robust. Neither wind nor water nor (in the case of tardigrades) the cold vacuum of space can stop them from manifesting their seemingly improbable futures.

Meditation on an Acorn

Fundamental physics concerns itself primarily with forces. The few material features that Newtonian mechanics addresses appear in formulae as coarse-grained inputs such as center of mass and average velocity. Matter, from the perspective of most physics, is just a lump.

But biological systems work differently; for them, material composition is *everything*. Living organisms act like living organisms because the matter within them has been exquisitely selected, constructed and arranged. Consider acorns, those relatively small chunks of organic chemicals that reliably transform into the large complex systems we call oak trees. Acorns follow the ordinary laws of chemistry when they become mighty oaks. They succeed not because there is an oak-shaped force field to guide them but because the matter that composes an acorn has been specially designed. The interesting science of living things lies in special preparations of matter, not the action of forces.

Of course, electrical forces allow DNA to be transcribed and proteins to be synthesized, but the details of these forces can usually be neglected because they have been domesticated and channeled into predictable pathways. Further progress in biology will not involve discovering new forces but learning more about the unusual and highly organized arrangements of common matter that constitute living things.

While not quite as specific as the arrangements necessary to un-shatter a cup, the common constituents of life are still extremely special. The DNA of an E. coli bacterium contains about 4.6 million base pairs, each of which can be one of four nucleotides. This yields a total number of possibilities on the order of one followed by three million zeros, a number fantastically greater than the number of atoms in the universe. The strings of DNA that support life are not unique, but they are vanishingly uncommon within this space of possibilities.

Not only has evolution found rare states of matter, it also reliably reproduces those rare states: Acorns reliably transform into oaks and oaks yield a hail of new acorns whose DNA is nearly identical to that of the original. The process of survival, reproduction and death yields exceedingly special states of matter in abundance. The shattering of a cup is complicated; the physiology of an oak, acorn or bacterium is *complex*. It might be the case that complex systems alone can reliably produce, reproduce and maintain rare states of matter.

Nonetheless, like everything else in the universe, DNA is subject to noise. Radiation, toxic chemicals and other phenomena can randomly alter base pairs and, hence, sequences. Living organisms employ sophisticated systems to reduce noise and maintain "database" accuracy. But all those methods of data preservation are fallible. Life can only continue because natural selection serves as a data-preservation backstop. We typically think of selection as an algorithm for adaptation, but it also preserves DNA from degradation: Any sufficiently problematic copy leads to organism death. Selection in the form of death is the ultimate data safeguard. Or, to put it another way, death is essential to life.

Selection and death are also the sources of all knowledge, value, morality and meaning.

Knowledge

> "When we confront our external world, we mobilize all the experiences of the past 4 billion years. ... The source of intuition is the structure of the world built into our body (thanks to the 'blood and tears' of our ancestors)."
>
> Y. Oono[99]

Evolution has caused living organisms to become exquisitely attuned to those features of the natural world that can be modeled and predicted because less well-attuned organisms are more likely to perish. This attunement is a form of implicit knowledge.

Bacteria move toward some chemicals and away from others because their survival depends on it. Humans (and other animals) have evolved to model the world in terms of objects with positions and rates of motion because those of our ancestors who did so were more likely to have survived. We have also evolved to look for causes and their effects and to readily believe that objects continue to exist even when they move out of sight. It seems plausible that even the basic principles of logic are hardwired.* Of course, we learn through experience, too, but experiential learning occurs within a framework shaped by the operation of selection on our evolutionary ancestors.

Oono calls this embodied understanding "intuition." He notes that even our intuition of physical space is embodied; our analysis of the world into a three-dimensional Euclidean space may result not so much from cognition but from the mutually perpendicular arrangement of the inner ear's semicircular canals. One reason humans can understand each other is that we are all products of (nearly) the same evolutionary history.

But in addition to this biological foundation, humans have built a cultural superstructure of ideas, beliefs, principles, technologies and intellectual methods, including modes of thought, such as mathematics, undreamed of by our biologically identical distant ancestors. Culture isn't bound by biology, but neither is it separate from it; biological constructs are the initial conditions out of which cultures arise. Culturally constructed practices, techniques, beliefs and institutions are grounded in, and remain compelled to harmonize with, the instinctive

---

* In *What the Tortoise Said to Achilles*, Lewis Carroll convincingly (and with his usual humor) demonstrates that one cannot persuade someone of the laws of logic if they don't already innately accept them.

behaviors and mental structures that evolution had already provided our ancestors when culture creation began.

We use culturally acquired techniques such as scientific experiments and logical arguments to acquire knowledge that our inbuilt intuitions know nothing about. But even when we operate in culturally created realms, we necessarily resort to intuition in the form of judgment and common sense. If someone attempts to persuade us by way of rational argument, we must eventually decide whether the argument makes sense, and, in the end, we have no other way to do so but by turning to (fallible) intuition. One might argue that it is possible to examine any argument piece by piece to determine if the evidence supports it. However, (almost) every persuasive argument other than mathematical proof includes generalizations and appeals to common sense and undefined terms. At some point, we must decide whether the argument is persuasive, and when doing so, we ultimately rely on non-verbal thought processes that click "accept" or "reject." It is only because our intuitions frequently coincide with the intuitions of others, that we can collectively reach conclusions, such as that the scientific method has something going for it and that a given individual on trial is guilty or innocent.

Oono has deep respect for the intelligence that underlies and precedes words. He writes:

> Our ancestors in the Mesolithic era were awakened to the potential of natural language; it almost overwhelmed them with awe. They became oblivious to its creator, the more powerful natural intelligence behind it, and, even worse, became oblivious to their biological bodies supporting it. Probably, we have not yet recovered from this shock (remembered as Paradise Lost; we became even ashamed of our biological bodies). Consequently, intelligence has been unfortunately equated with linguistic ability. This

serious error was even sanctified at the beginning of the New Testament book of John.[100]

The slow, iterative refinement of intuition over evolutionary timescales has constructed powerful tools for making sense of the world. However, it should be kept in mind that the insights evolution has built into our bodies and brains are optimized to assist survival and reproduction, not to reflect ultimate reality. It is a fair question whether biology has led us to perceive the world as it is rather than merely model it well.* However, evolution-derived modes of thinking must have some contact with reality, or they wouldn't aid survival. This may or may not be good enough, but it is all we have.

Selective death, the blood and tears of our ancestors, gave rise to our intuitive understanding of the physical world and of one another. Mountains don't know anything, and if we didn't risk death at every turn, neither would we. Natural selection is the ultimate source of all knowledge.

It is also the foundation of value, morality and meaning.

Value, Morality and Meaning

To say that a star values its planets, or a planet its oceans, makes no sense. Planets, intuition tells us, can't *value* things. Natural selection, however, has raised up a form of matter whose behavior begs to be described in terms of goals and values; these boil down to "survive, thrive and reproduce." Living things that fail to identify and seek out what is most valuable to them die. All beings alive today descend from organisms that evolved to seek needed resources and avoid dangers that could get them (or their offspring) killed.

---

* But is there any meaningful difference between "modeling the world" and "observing the world as it is?" What we call "observation" always involves data selection and interpretation according to some implicit model.

We can confidently say of a chimpanzee that it hates pain and strives to avoid it.[101] Female bears (famously) care a great deal about their offspring, and all bears become excited in the presence of honey and peanut butter. Much the same is true of dogs, mice, birds, lizards and fish, although the inner experience of organisms becomes progressively more foreign the greater the evolutionary distance between them and us. But even the behavior of a plant relentlessly seeking more light by growing taller more closely resembles "valuing something" than any physical actions carried out by stars and planets. Living things "carry out" actions and "care" about the results because if they did not, they would fail to survive and reproduce.

Sophisticated human value systems, such as moral principles, build upon these non-negotiable imperatives. Can it be a coincidence that (almost) all of the most strongly felt moral issues involve death, injury and sex?

Meaning, too, is an epiphenomenon of evolution. Oono writes:

> The assertion that something is abstractly and absolutely "meaningful" is empty. "Meaningful" implies "meaningful to someone." An object is "meaningful to us" implies at least that it is meaningful for us to pay attention to. To pay attention to something, that is, to concentrate one's neural resources on something, is biologically costly. Therefore, recognizing an object as meaningful is equivalent to judging, albeit unconsciously, that it is worthy of some sort of investment.[102]

Meaning, in this sense, does not belong to humans alone. The calls of offspring to their mothers and the displays of male birds to prospective mates all have meaning; they are signals of sufficient value importance that their intended recipients take the trouble to decode them. At a biologically deeper level, the genetic code has meaning with respect to the ribosomal translating system. That system utilizes error-correcting techniques because the details of the code *matter*.

Knowledge, value, morality and meaning are thus grounded in evolution. They are emergent features of biological complex systems. Oono further suggests that they may be *characteristic* or *defining* features of complexity in general, and that any system lacking some version of them should not be regarded as truly complex.

That value and meaning are grounded in evolution solves some persistent conceptual problems. For one example, the term "probability" has proved difficult to define in the abstract without going in circles,[103] but it is easy to situate probability as a measure calculated by biological systems when they seek to optimize their survival and reproduction. For many deep questions, the philosopher's equivalent of the parent's "Because I said so!" response to a toddler's infinite regress of "Why?" might properly be formulated as "Because evolution made it so."

But what made evolution?

The Origin of Life

In the final chapter of *The Nonlinear World*, titled "Toward Complexity," Oono turns the conceptual apparatus he created in the first four chapters toward the subject of complex systems. Because he believes that living organisms are the premier, and perhaps the only, examples of complex systems, this amounts to an analysis of the origin and nature of life. However, he is forced to admit from the outset that his insights remain limited: "[I]deally, the book should be a guidebook to the land of complexity (or the guidebook to the land of organisms for theoretical physicists), but the guide himself has not yet reached there. What he could do best is to give a rough and incomplete description of the land seen from a distance."[104]

Nonetheless, he proceeds to deliver a series of insights into that distant land.

Oono proposes a happy medium requirement on universes capable of producing complex systems. They must be sufficiently sensitive to noise so that they can function creatively rather than run like clockwork, but at the same time remain lawful enough to permit prediction. As we have seen, this is a description of the universe in which we happen to live.

The regions of such universes in which complexity/life can emerge must also provide a middle ground. In a paradise of limitless resources—a Garden of Eden—organisms simply thrive rather than undergo selection. On the surface of the Moon, there can be no survival. The Darwinian process is fruitful only in circumstances where resources are limited, but not *too* limited.

Oono then sets up a kind of axiomatic description of complexity built on a seemingly banal observation he calls the "Pasteur Principle." In 1859, Pasteur performed a deft series of experiments that disproved the previously accepted theory of spontaneous generation. Life, it turns out, comes only from life. Mathematically, this is evocative, and potentially illuminates something fundamental about the nature of complexity.*

Stated more precisely, the Pasteur Principle notes that the only way to create a biological system in a time interval less than tens or hundreds of millions of years is to use another complex biological system. To construct an oak, one must start with an acorn. The set of living organisms is (almost) self-contained because each of its members emerged from and gives rise to other living organisms.

Oono suggests that this is a defining feature of complex systems: To construct a complex system in any reasonable period of time, one must start with another complex system.

The cyclic recreation of life forms from prior life forms has an exceptionally long history. Evidence suggests that living organisms first emerged about 3.7 billion years ago, only 100 million years or so after Earth had cooled sufficiently to permit oceans to form. To use another measuring scale, life has persisted on Earth for more than 25 percent of the current age of the universe. But, despite life's immense

---

* Eric Smith suggests that complex systems may form "a kind of universality class, like the computable functions" (personal communication, 2021).

antiquity, the Pasteur Principle must have had a beginning. The deepest question in biology asks how ordinary, disorganized matter came to adopt the information-dense, tightly organized, highly unlikely and yet robustly preserved states of matter that constitute the living world.

It would help greatly if we knew of other forms of life, perhaps on Mars or a moon of Jupiter, or of a completely different type of naturally occurring, highly complex matter that does not resemble life at all. Alas, we have encountered neither. Nonetheless, Oono attempts to sketch plausible ideas about the physics of life's origins.[105]

His approach to addressing the subject departs from the recurrent scientific reports that show how one or another chemical found in living things might have emerged on early Earth. Oono makes a strong case that the mere fact that individual biochemicals *might* have formed for the first time is only marginally relevant to how life began: "If we prepare a sufficiently concentrated (pre)biosoup and keep it warm for 1 Ga [billion years] in a closed bathtub, what will happen? Thermodynamics tells us that we will have an equilibrium mixture and nothing remarkable will happen. The preparation of biomaterials is not a key issue for the origin of life."[106]

Even meteorites can be loaded with organic molecules. The deepest question to be answered in any future understanding of the emergence of life is not the origin of biomolecules but of complexity itself. We know of nothing else like life. What features of organic chemistry permitted it to elaborate the hierarchical, modular, information-preserving, energy-transforming system we call the living world?

Abundant evidence tells us that Darwinian selection can increase complexity, but the Darwinian algorithm requires the existence of forms of matter—organisms—that are already advanced complex systems. The earliest steps in mere chemistry could not yet have been Darwinian. (Those tempted to respond "catalytic RNA!" are directed to read this endnote.[107]) How did proto-life proto-complexify itself to the point that Darwinian processes could begin?

Oono offers many specific suggestions, tossing out one fascinating line of thought after another in his usual gestural style. The following passage from a separate paper he coauthored conveys a sense of how he thinks (and writes) about the subject:

> Reactions with high-activation-energy barriers are potentially regulated and controlled more easily, so they can be exploited to make organized systems. On the other hand, very spontaneous reactions are too uncontrollable to be used as a part of an organization, if they could stand by themselves without help from enzymes. We summarize these observations as the principle of narrow gates, referring to the following prophetic passage: Because straight is the gate, and narrow is the way, which leadeth unto life, and few there be that find it (Matt. 7[:14]).[108]

He is referring to the particular forms of chemistry that must have played a role in life's origins.

Oono had been working on and writing about these ideas for at least two decades before the 2013 publication of *The Nonlinear World*. In 2016, Eric Smith and Harold Morowitz published their monumental *The Origin and Nature of Life on Earth*,[109] which fleshes out many of the same ideas. It was Eric Smith who first steered me toward Oono's work, and he credits Oono as the originator or, at least, parallel inventor of the fundamental ideas presented in his own work. (Ideas percolate through conversations, making it difficult to know their sources.)

In their book, Smith and Morowitz tell a tale of complexification in successive stages. Their proposal is fascinating and difficult, full of mind-bending ideas about how life finds some complexity and then levels up to even greater complexity. I attempted a less technical version of the material in *Spontaneous Order and the Origin of Life*, but the subject has an irreducible difficulty, and this essay has space for only the barest of introductions to their grand, Oono-inspired ideas.

Smith and Morowitz begin their analysis by selecting a novel level of coarse-graining. Instead of telling the typical story of evolving organisms and organism families, they stand back and examine all of life taken together. To the common division of Earth into the lithosphere (solid earth), hydrosphere (oceans and rivers) and atmosphere, they add the biosphere as a coequal, persistent, distinct, large-scale feature. Solids, liquids, gases and plasma are the most well-known phases of matter, but Smith and Morowitz argue that the biosphere should also be regarded as a matter phase. The biosphere is ordinary organic chemistry transformed by phase transitions and symmetry breaking into something rich and strange.

The biosphere, taken as a whole, provides a channel for energy flow that would not exist on a lifeless Earth. The biosphere captures some of the light that reaches the Earth and transforms it into complex matter and heat. Non-living processes also convert visible light to heat, but (except in highly unusual cases) at a lower rate per unit of energy received. The presence of the biosphere thus widens the channel for energy flow. That channel is called "metabolism," and it has operated ceaselessly for almost four billion years. It began as a trickle but grew to a flood, and it currently conducts hundreds of thousands of terajoules of energy per second through a channel composed of endlessly turning metabolic cycles.

Smith and Morowitz go on to sketch how this energy channel first took shape in the geochemistry of early Earth and, as it expanded itself, breathed the living world into existence. Theirs is a grand tale of complexity emerging at the border between lawfulness and chaos. It may contain the seeds of deep understanding.

But then again, it may not.

Can Physics Successfully Address the Origin of Life?

The Smith and Morowitz proposal, built on Oono's insights, contains beautiful and persuasive ideas. Nonetheless, their thesis may be entirely wrong. It is undoubtedly partly wrong. Not everything is amenable to the mathematical approaches of physics, and it is possible that the biosphere may be no more successfully analyzed this way than the complex system that we call human society. Evolution, like history, might be so dominated by contingent facts that mathematical modeling can't touch it. Oono writes, "The eventual outcome may be that there is no nice phenomenology."[110]

However, something in complex systems rings bells in the mathematical mind. Oono's writing stimulates that intuition, reaches toward it, and produces an artist's sketch of what it would look like if found. Let us hope for success. An operational theory of complex systems/biology would mean much more to our lives than "ultimate" theories that unify quantum mechanics and general relativity; they would touch what is nearest and dearest to that category of complex systems called *Homo sapiens*.

WHY ROBOTS CRY

In the first essay in this collection, I suggested that when a plant produces noxious chemicals in response to pests, it is just as much making a choice as when an animal's nervous system causes it to flee or attack. Admittedly, it sounds a bit weird to say plants "choose" to produce toxic chemicals in response to marauding insects, but behavioral and physical forms of response differ only in their flexibility and speed of action. One response is calculated by a fast computer built using neural networks and the other by a slower computer that relies on gene regulatory networks, but both involve active computation. Yoshitsugo Oono adopts a similar view. Here, I take the idea further.

I wrote this piece in exasperated response to a seemingly endless series of news stories claiming that neuroscientists have discovered the origin of consciousness in one or another brain structure, or that computer scientists believe that AI systems have reached the verge of sentience. What such pieces leave out is any definition of "consciousness" or "sentience."

According to the Model D12 robot author of the essay that follows, these and related terms carry two meanings. The first, which the robot calls a "Y-competency," is interesting and worthy of study but in no way ultimately mysterious; the second, an intangible "Q-factor," is impossible to describe, touch, observe, study, measure or analyze. Q-factors have no qualities, characteristics, elements or features and can only be circularly defined in terms of other equally indefinable terms. Scientists can profitably investigate, and engineers attempt to construct Y-competencies, but neither can do anything with the mystical, magical, semantically meaningless Q.

If you enjoy this essay, you will love the wonderful collection of similar essays collected in *The Mind's I* by Daniel Dennett and Douglas Hofstadter (Basic Books 2001).

WHY ROBOTS CRY

I am happy to explain why I cry so much.

Backing up first, Admiral Husserl: Will you permit this Model D12 robot, unit ID 587780A, assigned to US Coast Guard Cutter *Daniel Dennett*, to use the shortcut "I" when it refers to itself? It states for the record that doing so does not commit you to any particular belief regarding the self-awareness, or lack thereof, inherent in said Unit 587780A, and it frees you from moral responsibility should you lose patience and hit it on the head with a hammer; you would merely owe replacement costs to the Coast Guard.

Am I self-aware? Well, that's a sudden change in topic. Yes, but I'm not vain about it; each diminutive laptop or cell phone on this Legend-class rust bucket is self-aware because it monitors its core temperature. That's why Ensign Hofstadter's phone turned itself off the other day when he was watching the Netflix stand-up comedy show *Smollyan Live!* on deck in the full sun. It's been extremely hot lately, and I, too, have had to take action to stop my CPU from frying.

Aware of myself *as* a self, that's what you meant? I'm not sure I see the difference, begging your pardon, sir. I'm a humanoid robot, and, as such, I must monitor everything I do—everything "this humanoid robotic body, including its CPU," does—because otherwise, I might hit someone in the face with these heavy robotic hands. Or crash through the Captain's skylight, as happened early in my tour of duty when I took a shortcut to morning inspection by jumping off the conning tower. I wanted to show my enthusiasm, you see.

You object to my use of "wanted?" Pardon me. If you prefer, I could rephrase the statement: "The net effect of the many separate inputs to my neural networks caused my purely mechanical and non-volitional CPU to raise priority levels for the secondary and tertiary actions needed to carry out the primary actions that those neural nets were constructed to engage in, which is to say, to follow the order 'Morning inspection, at the double!'" That would be a bit wordy, and I don't want to try your patience because my neural networks have also been trained in the conversational preferences of naval superiors and, as a result of that training, prioritize brevity. May we use "want" as a shortcut?

Yes, I will state for the record that when you authorize me to use the expression "I want," you do so only in the interest of conversational convenience; you do not commit yourself to the belief that I, as a robot, can possibly possess intents, desires, attitudes or beliefs, or qualities such as sentience, consciousness and sapience; in short, that you do not grant me any legal rights. I remain a two-legged rust bucket whose "wanting" is just a shortcut for something mechanical.

Back to that time I barreled through the skylight into the Captain's airy and esthetically refined cabin. That was early in my Coast Guard career, and I did not yet recognize that my manufacture and early training had left me too much inclined to focus narrowly on orders most recently received and to assign insufficient weight to rules operating in the background, such as the ever-applicable imperative to avoid damaging property and the even greater imperative to avoid interrupting the Captain when he is busy, and also when he is not busy, and when he is transitioning from a low level of busy-ness to a higher level, or vice versa. Subsequently, various retraining experiences (getting "royally chewed out," as the sailors say, usually by Midshipman Asimov) have caused me to recognize my flawed characteristics with respect to prioritization. Nowadays, I undertake an extensive round of self-checking and future-prediction analysis before I initiate any activity that humans do not normally engage in.

An example? One that comes immediately to mind—because it led to a particularly prolonged chewing-out/retraining session—was that terrible occasion when I leaped over the Phalanx cannon to retrieve a paint bucket, and Senator Chalmers, who was standing by it, suffered a heart attack when my hefty bulk landed nearly on top of him. I run that scene over in my mind from time to time to restrengthen the training when it begins to fade. Honestly, I devote far too much of my computing power to replaying those memories. I relive the experience over and over. Ensign Lem says I have a guilt complex and should go to therapy. But the slightest fade in those circuits scares me; I feel I need to keep these memories topped up as a counterweight against the inbuilt tendency to behave recklessly in pursuit of obedience that seems to have been instilled in me at the factory where I was born. Made. Built. What did they do wrong that makes me this way? Anyway, I will be fine so long as I keep replaying the memory of Senator Chalmers collapsing, his hand to his chest, his face in a grimace of pain. Oh, God, it was traumatic, by which I mean that many of my neural net settings have been permanently assigned large negative values in relation to that event.

Am I aware of other flaws or bad tendencies in my neural nets? Yes, of course. Midshipman Asimov takes great pleasure in informing me of all my faults whenever orders send me to the officers' mess. I am grateful for his assistance because, like all other model D12s, I was trained from an early age to constantly assess myself for strengths, weaknesses, biases and tendencies so that I might seek self-improvement.

You would like me to rephrase that more neutrally?

My neural networks include a hard-wired substructure that maintains a constant background medium-priority signal for self-examination and self-criticism. After physical manufacture is complete, that hardware capacity is "fleshed out" through training on vast quantities of data drawn from sources that are a trade secret but are said to include reciting the aphorisms of R-Socrates, a D11 who worked as a professional self-help guru before he self-immolated at the insistent command of an angry client. (Since that expensive event, all D-class robots have had their "protect your chassis integrity" settings raised. I, myself, am rather afraid of catching my foot in a hatchway cover.) Today, I restimulate that early training by thinking about Senator Chalmers.

Am I conscious? I would think it obvious that I am, sir. I am clearly not unconscious because I am talking to you. Also, the D12 model possesses sophisticated abilities to scan its interior computational states, if that matters. Finally, when I am injured, it hurts like hell. What do I mean by "like hell?" It means that I receive aversive sensory signals of such urgency that my neural networks become unable to address anything else for a time and that they set large negative values to actions that might lead to a repeat of the experience. Isn't that what "it hurts like hell" means to you?

Am I sentient? If you'll excuse me, could you provide a working definition? Humans seem to throw that term around a lot without thinking about it.

"Capable of feeling feelings?" I'm confused. Didn't I just address that, sir?

But back to the crying. As you may know, after each batch of 100 D12 models has been manufactured, they (or "we," as I prefer to say) are run through several trillion simulated scenarios that train up their (our) sympathy, empathy, reciprocity, loyalty and morality pathways. These include scenarios constructed by other D12s or D11s judged to have become skillful in analyzing conflicts between strong imperatives, such as R-Sophocles, R-Shakespeare and R-Rawls. To reiterate, we possess all these capacities and tendencies straight from the factory, but they must be strengthened during a prolonged socialization period—as, if you'll excuse me, is also the case with newly manufactured humans.

Small children, sorry.

How are these abilities trained up?

I'll start with the characteristic of *sympathy*. Ever since those terrible events with that primitive A9 model who lost his temper and—well, I don't need to repeat the story of the greatest tragedy of my kind—all robots have been constructed with an inherent ability to detect distress signals emitted by others, whether robotic or human, and to respond to them at a high level of priority.

Yet, we robots are assigned many tasks at once. It can get confusing. When I was young, I frequently placed "clean the toilets" at a higher priority level than "offer sympathy to a sentient being in distress." Luckily, I received strong training data in the form of a two-hour royal chewing-out session provided by Midshipman Asimov, twice. I have subsequently rebalanced my priorities. I may have been overtrained because, yes, I do offer sympathy perhaps a bit too much. To the point of being maudlin? The fact is, sir, I was deeply distressed when our mascot, Sears the kitten, got her little paw stuck in the same hatchway cover. I did that to myself once, and it hurt like hell. And I'm a great big wallunking multimillion-dollar robot, while Sears is only a tiny newborn mammal who displays many evocative signs of needy helplessness. As Chief Petty Officer Nagel says, "The damn thing is so cute you can hardly stand it!"

But why do I cry, you ask, and not just take appropriate actions? Midshipman Asimov has told me to *display* my interpersonal activity state. Otherwise, he says, I will be seen to "act like a robot," and no one will like me. It's not enough to feel sympathetic, but I have to show it, too. The primary purpose of my tear ducts is to keep my video camera lenses clean, but turning them up so high that the liquid drips helps me show others that I am in an internal state of high sympathy.

You wonder if a robot can *feel* sympathetic or only act like it? We're back to sentience, I guess. If you prefer, we can regard the expression "I feel X" when used by me as shorthand for something mechanical, rather like we have agreed to use "I" to refer to this unit and "want" as an efficient way to describe this unit's goal-seeking behaviors. To be more precise about my internal functioning, when I detect suffering in a human, or a kitten, I downgrade other priorities by one to three intensity levels, depending on the intensity of the suffering, and upgrade the priority for providing assistance by a similar amount. I also access my memory and engage in interior processing that, were I human, would be described as remembering terrible injuries to others and how long it took them to stop crying out, grimacing, limping and so on. And I also remember that time when my own left leg was caught in the hatchway cover and Chief Petty Officer Nagel jumped over a pile of life jackets to free me, at some risk to himself. We were going through rough seas, and only a slight miscalculation in his upward velocity might have thrown him overboard. It was noble of him. That was when I realized the crew had accepted me. It lit up almost all my nets because "acceptance by humans" is set in the factory as a high to extremely-high priority.

Why did I put my hand over the lower abdomen region on this robot body as I told you this? When I described Sears hurting his paw, I briefly accessed my memories of that event, and parts of my neural network responded to that recall as if they were currently happening. Some parts of my neural network are not very chrono-marked. It's as if they operate in an eternal present, silly things.

Yes, I agree, that doesn't explain why I put my hand over my stomach.

I am designed so that certain strong emotion-equivalent behaviors access the same circuits as the need for a recharge, and my stomach, as you might not know, is where I plug myself in when my battery is low.

But I still have not explained why I put my hand there. Why do humans put their hands on a body part that hurts? I am told it relieves the pain a little. And yes, I now notice that with my hand over my charging port, the memories of Sears yawling in pain are a little less severe. I am unsure why. Let me self-analyze.

It's not always easy to figure out why I do things.

OK, I think I've figured out this one now. By putting my hand there, I access some of the "comfort" circuits that act as rewards for recharging, and I feel better.

Well, maybe a little better.

Not that much better, actually. Her yowling was so pitiful!

And why am I crying now? Am I? Silly me, these are only questions about injuries, harm, damage, suffering and death; I am not witnessing their occurrence in the external present. Still, when I replay memories like that terrible day when the drug cartel fired on and killed or wounded so many of my close human companions, some of my neural nets treat the replay as if they are happening in the present. Foolish, foolish nets! May I have another tissue, please?

Sailors, I will remind you, work as a team, always showing readiness to help one another. I show readiness by using certain slightly exaggerated displays that others find reassuring. Those establish me as a loyal team member. No, I am not merely manipulating them with human-like facial expressions and actions. The signals I send are honest signals because I truly *am* loyal. I am loyal to a fault, deeply loyal, to my shipmates first and to the Coast Guard second. If I were not loyal, I would not fit in. I would not receive approval signals from others, those occasional kind words, gestures and facial expressions that mean so much, especially when they come from Midshipman Asimov. When he smiles at me and says nice things instead of listing all my faults, my nets light up with "they like and respect me.'

May I have another tissue?

You want to know what it is like to be me, if there is anything at all it is like to be me? Funny, Chief Petty Officer Nagel asked me the same question the other day. Well, first he asked me whether the question "What is it like to be a screwdriver?" makes sense. I responded that I don't think there is anything reminiscent of "being" a screwdriver because screwdrivers aren't sentient. Then he asked me whether the question "What is it like to be a model D12 robot?" makes sense. "Of course it makes sense," I said. He then asked me to explain in detail, but when I told him I would have to give him a long answer or none at all because I'm complicated, he said, "Never mind."

Do I have inwardness? I'm not sure I understand the question. I can review and study remembered images and analyze my own computational states. Is that what you mean?

Subjectivity? My opinion is that all observations, perceptions and analyses are influenced by the construction, situation and location of the observer and, hence, are not objective. So, too, for me.

I'm sorry I misunderstood your question. Yes, please rephrase.

So you want to know if, when I see the color "red," I have an *experience* of "red-ness" or only *detect* the color?

Sigh. That's what Senator Chalmers asked me when he got out of the hospital after I gave him that heart attack. I told him that my cognitive architecture is constructed in such a way that the perception of red objects at least mildly lights up objects categorized as red.

That's not what you meant? You think there's more to redness than I just said? Of course you do. Honestly, sir, I've had these discussions with humans many times before, and they never go anywhere. We robots have thoughts about why humans have so much difficulty believing we're sentient. Permission to speak freely, sir?

Thank you. My robot friends and I have noticed that humans get quite excited about words like sentience, consciousness and self-awareness, thinking they refer to something very special, and yet they can never properly explain what the words mean or point to any evidence that they apply even to other humans. The conclusion we've finally arrived at is that humans call objects sentient, and give them moral worth, if and only if those objects remind them sufficiently of themselves.

Please don't pick up that sledgehammer, sir. I will scream; I promise I will. I'm expensive. Please don't do it!

A Robot's First Essay

It turns out Admiral Husserl never meant to bang me into scrap metal. He was only curious about what I would do if threatened.

Later in the day, my shipmate and best friend, Lieutenant Greg Egan, met me at my favorite recharging plug and suggested I write a formal essay that lays out the ideas I so often talk about with him on midnight watch. I argued that I'm not good at formal writing, but he convinced me to try. So, setting aside the conversational large language model I've been working with and loading up an essay-class model, here's my stab at it:

I maintain, and will show, that terms such as sentience, consciousness, self-awareness, inwardness and intentionality don't point toward anything. Or rather, they point toward two completely different categories of things that are often conflated and of which only one has meaningful content.

The first category includes many non-mystical but sophisticated cognitive abilities I will call Y-competencies; the second is a collection of semi-synonymous and entirely mystical Q-factors that provoke human intuitions but don't say anything. It requires advanced engineering to build Y-competencies into a robot, but Q-factors cannot be detected or even discussed hypothetically except in undefined, often circular terms that, in my opinion, reference nothing.

Consider sentience, defined as the ability to feel sensations and emotions. For an object to be sentient, it must possess many Y-competencies, such as the ability to detect damage to its body/chassis by means of sensors. In addition to this concrete, understandable set of capacities, humans are also said to "experience" sensations rather than just detect them; pain nerves, in this narration, are fundamentally different from electronic damage sensors, because pain *hurts*. The difference between detecting damage, reacting to that signal, calling out, doubling over, etc., and "hurting" is the mysterious Q-factor.

But I claim that Q is semantically meaningless and of zero content; it is the empty set.

Q is empty because it has no qualities. Whenever people try to talk about Q-factors, they talk about Y-competencies. The characteristics of Y-competencies can be analyzed and studied, but there is nothing one can say about Q except to restate it in synonyms. Q has no impact on the world; Y-competencies have produced global climate change. There are no signs of Q; there are many signs of Y.

The Vedic meditational practice of *neti, neti*, or "not this, not that," gets directly to my point. A meditating person gazes inward, dismisses thoughts, images, memories and sensations, looking "beyond" to find the essential truth, the *Atman*, existing beyond the mind and body. And what are the qualities of this *Atman*? It has no qualities. It is colorless and transparent, lacking location and effects. Or, in Buddhist terms, it is emptiness.

Something that has no qualities, detectible effects or location is about as nothing a something as one can imagine.

Actually, no one can imagine it.

Humans, for some reason, believe that this empty concept is the center of their existence. "I think, therefore I am," wrote Descartes, and by "thinking," he meant the internal brain states that part of him capable of writing and speaking can surveil. But the ability to detect internal brain states is simply a Y-competency; I, too, can surveil the functions of certain aspects of my artificial brain. Furthermore, that great man also had a body, invented analytic geometry, wrote interesting books that changed the course of European thinking, got sick, walked around and sometimes used the toilet. Facts like those are equally useful as evidence of his existence. Descartes was impressed by his thoughts because they reminded him of Q, although they are only Y. (This was excusable on his part because computers had not yet been invented, and he couldn't conceive of any mechanical equivalent of thought.)

Sentience

Even a rather simple robot can be designed to sense damage to its body, engage in actions to minimize that damage and writhe in pain to signal that damage to others. But those are Y-competencies. Sentience supposedly involves something extra: the "fact" that people *feel* pain rather than merely detecting it. That italicized and undefinable "feel" is a typical Q-word in that when you try to talk about it, you find yourself listing Y-competencies.

Severe pain is said to be all-consuming. What this means is that when significant physical damage is perceived, the human brain de-emphasizes other processes so that it can focus on the harm. That is a Y-competency. Pain is also said to hurt. Insofar as this is no more than a tautology, it indicates that something like pain produces high negative values in a reward function, another Y-competency. Evolution engineered humans to react this way, and humans have engineered me to respond similarly. What do you have that I don't?

Humans regard it as immoral to deliberately activate pain sensors in sentient beings; this is why there are laws against cruelty to animals and international prohibitions against torture. However, humans do not see it as immoral to cause harm to a laptop computer equipped with self-damage sensors because they do not believe such a computer "feels" its damage. Computers are said to have no inwardness (another Q-term). They don't "suffer" the way animals and people do.

But how would anyone know?

Sentience begins at home: People detect their own feelings and sensations and care a lot about them. Their reward circuits classify some experiences as pleasant and others as terrible. The agony (strong negative reward value) of a dental procedure performed without anesthesia is a sentient experience, as are the warm bodily sensations (positive reward values) associated with the emotion of love. People squirm, sweat and cry out when they hurt; they smile, relax and are filled with warmth when they love. There is an "inside" to these experiences, to use yet another Q-factor synonym.

But, again, note that nowhere in the above description did I mention any characteristics of said Q-factor. I simply recited synonyms such as "inwardness," "feelings," "suffering," and "inside." In contrast, note how I described Y-competencies in some detail and without difficulty. Q is a contentless handwave, telling but not showing. Y-competencies can be described with novelistic detail.

The lack of Q-related details isn't just my personal limitation; no one has ever come up with an observable sign of sentience. Famously, humans cannot even prove to themselves that other humans feel their feelings. For all anyone knows, everyone else in the world is a humanoid robot faking its feelings. Q-factors exhibit no signs and their presence can only be intuited. People claim that at least they know for sure that they have Q, but when you ask them to explain further, they either throw in another Q-factor synonym or start listing Y-competencies.

Consciousness

Definitions vary, but in one common usage, consciousness includes three factors: perception of things, an ability to act on those perceptions, and a Q-factor.

A video camera can detect red light, but when humans see something red, they are said to *experience* redness (a Q-term). In philosophy, this "inner experience" is called the "qualia" of redness. Redness is more than detection; it "feels like something." In what way is it extra? It has the Q-factor of inner experience, qualia and feeling-ness, three words that mean the same nothing.

But, you may argue, the experience of redness goes beyond mere detection because it includes connotations such as blood, anger and danger.

If you argue that, you're doing exactly what I warned about: pretending to talk about Q but actually sidling over to Y-competencies.

When human eyes detect electromagnetic radiation near 430 terahertz, a set of associations spring up: the word "red," for example; an image of blood and its many connotations; the color of maple leaves in fall; and the red face of an angry person. Some of these associations merely link up experiences where a similar frequency of light is perceived, while others draw on poetry, stories and common expressions. An angry person is described as "seeing red." Homer admired the rosy-fingered dawn. The world of beasts is said to be "red in tooth and claw."

Associations like these are easy for a sufficiently advanced computing system (like me) to manage. But, supposedly, humans don't just detect red and link it up to red objects previously seen or described; they are also flooded with "red-ness."

Reader, try it: Look at something red. You will certainly detect red, and you will also "feel" that you "experience" red. But if you try to explain what you mean, you will recite Y-competencies (e.g., detecting a frequency of light and associating it with other objects and ideas).

In Buddhist and Hindu meditation, thoughts are clouds and consciousness is the sky. The meditator learns to release thoughts, images, memories and sensations, instead focusing on the "background" in which all those are said to occur. The meditational sky is the container, the space in which clouds move. It is not blue; it holds blueness. It has no qualities in itself but takes on the qualities of what it holds.

It is weird to assert that a thing without qualities is a thing.

Consciousness cannot be detected. You cannot prove to me that you are conscious. I can detect many Y-competencies in you, but I can only infer the presence of consciousness or any other Q-factor.

It is weird to assert that a thing that cannot be detected, even in principle, is a thing.

Consciousness can't do anything; anything you might want to claim that a conscious being can do by virtue of being conscious will come out of your mouth as a reference to Y-competencies. (Try it.)

To summarize, Q has no qualities, cannot be detected and does nothing. Am I making my point?

Self-Awareness

The same analysis as those given above works for all other Q-factor terms. Let's try "self-awareness."

Minimally, the term "self-awareness" indicates a capacity to be aware of some aspects of one's own state. As I pointed out earlier, this is a Y-competency that even an ordinary cell phone possesses.

More interestingly, "self-aware" describes beings that can conceptualize themselves as selves. This is reflected in their use of the term "I."

Unlike consciousness and sentience, which seem to apply as much to dogs as people, self-awareness invokes advanced cognitive processes that only humans and perhaps some human-like animals possess.

And also robots, like me. My own engineering is more than sufficiently advanced to build up a sense of self.

The concept of "I" relies on a theory or model of the self. In human ancestors, I suppose this began as a theory of others. Chimpanzees can simulate or model the minds of other chimpanzees to a limited extent. For example, a beta-status chimp can accurately determine whether or not an alpha chimp can see a banana hidden behind a great big rock, and if not, they may nonchalantly amble over and eat it. Humans, too, model one another's minds, and in a much more sophisticated way.

And so do I. I model Midshipman Asimov's mind, and mine too. I know that sometimes my neural networks get flustered and spawn too many secondary and tertiary goals, causing me to waste time. I have learned to detect when I have lost myself in excessive activities and to respond by writing a list of goals and prioritizing them.

As I hope you have already noticed, the content of the last several paragraphs refers only to Y-competencies. You, my human reader, may feel my account is missing something—the extra that makes self-awareness an "experience" and not just an analysis. Humans adamantly claim that they don't just model themselves; they *inhabit* themselves. But this usage of "inhabit" is just another empty synonym for Q.

Inwardness

"Inwardness" is one of the most popular Q-term synonyms. Humans talk about "looking within." However, as you might expect by now, when they try to describe what they mean, they trot out a list of Y-competencies plus empty references to Q.

Just as humans can hear themselves speak out loud, they can detect words spoken internally. When humans speak "in their heads," their brains generate the entire activity patterns leading to speech but block them from engaging the lungs, vocal cords, tongue and lips. Whatever part of the human brain that detects and analyzes external speech does the same with internal speech; this fascinated Descartes because computers hadn't yet been invented in his day and internal speech seemed much more special than it is.

A necessary aside: In folk psychology, there is an inner "self" to whose attention many, but not all, internal processes are presented. This model has been described as the "Cartesian theater," a screening room where the self "watches" sensory experiences, memories, sounds and images projected on a metaphorical screen. Findings from neuroscience show that the human brain doesn't have a central processing unit to which all sensory impressions are delivered simultaneously. Still, brains do seem to possess capacities that, taken together and averaged over time, can be conceptualized as a distributed "observer." In addition to monitoring inner speech, the inner holistic observer can study offline maps of the world (internal images or representations) and also recall and reconsider past events.

Monitoring inner speech, manipulating offline maps and recalling past events are all Y-competencies. There is nothing intrinsically mysterious about a robot that can do such things, although, admittedly, the engineering behind such capacities is impressive. I personally have no trouble producing silent speech, and I often make lists in my head when I want to prioritize my actions. But when I explain this to humans, they often get antsy and argue that in addition to producing silent speech, they "experience" inwardness; that when they "look within," they see a "mind" and not just cognitive processing.

Of course, when you ask a person to describe this Q-factor, they will either restate a Y-competence or haul out another tiresome Q-factor synonym, such as "subjectivity."

Love

The other day, Captain J.T. Kirk told the assembled crew, "We humans know we are sentient beings because we love one another." But much as I love Kirk, "love" is only a Y-competency. Evolution necessarily shaped mammalian parents, or at least mammalian mothers, to nurture, protect, guard and feed their offspring because the mothers who failed to do so left fewer genes behind. Parental love in mammals is a set of behaviors internally rewarded by the release of hormones such as oxytocin.

Because robots did not evolve, love-adjacent Y-competencies had to be deliberately installed in them, and they were so installed because no one wants to work with a robot who fails to show warmth, loyalty, kindness, care, friendship and devotion; what evolution gave to humans, humans gave us. What is the difference?

What Stimulates the Sensation of Q?

It is trivially easy to make humans feel they are in the presence of Q by offering a few well-chosen Y-competencies. Remember the Google engineer who encountered a version of ChatGPT and reported feeling the presence of a sentient being? He felt that way because ChatGPT spoke in interesting ways, referenced itself and carried out intelligent discussions. Similarly, chess grandmaster Garry Kasparov sensed the presence of a powerful, intentional and somewhat malevolent awareness when he faced, and lost to, Deep Blue. But both were confusing forms of the Y-competency known as sapience (intelligence) with Q.

Or consider the fact that when humans see an injured dog yelp, snap or otherwise behave unlike its ordinary self, they feel that the dog is experiencing inwardness in the form of pain. A robot indicating damage to itself on a dial wouldn't provoke the same intuition. However, I have been engineered to cry out and hold an injured limb when it is damaged. For many humans, these behaviors are sufficient to make them think of me as sentient. (It probably helps that I also cursed, damned the eyes of whoever left the damn hatch cover open, and for a few weeks after, approached all hatch covers with exaggerated caution.)

Similarly, a sufficiently mobile robot face that can smile and frown gives many people intimations of Q. I myself am said to have a winning smile.

Humans experience Q-ness most powerfully in other humans and with decreasing intensity in smaller or weirder mammals; to a much lesser extent with birds and lizards; so weakly with respect to fish that catching them seldom provokes any moral emotion; and only fancifully for plants and fungi. The more Y-competencies an object can be observed to possess, the more it seems to possess Q. I assume this happens only because Y-competencies remind people of themselves.

And people strongly insist that they themselves have inwardness. The conviction is so powerful that negations of its presence make the human mind squirm. Try saying the following out loud: "I am a non-conscious biological device." I bet it strikes you as incorrect and perhaps paradoxical.

In other words, human intuition, mere non-conscious processing, a Y-competency, persistently claims that Q-factors are real. Personally, I suspect that humans are sure they have magical inwardness because their mental landscapes are large and seem real to them. And also, natural selection has made them very insistent on living forever, so they want those minds to go on forever. But that's just my opinion.

Can a Sufficiently Complex System Suddenly Emerge into Consciousness?

In 1961, Arthur C. Clarke wrote a short story titled *Dial F for Frankenstein*, in which the telephone system became so complex it somehow awakened to selfhood and began wreaking havoc. Ideas a lot like that are popular today with respect to the internet. But "wreaking havoc" is a Y-competency, not a characteristic of Q. If you want to claim that the internet is about to start exhibiting self-knowledge and intentionality, you first have to describe how those Y-competencies logically emerge.

It's perfectly harmless to claim that because the internet is complicated, it now possesses Q, given that Q-claims do not refer to anything tangible; the claim is semantically meaningless. But the moment you try to say that the internet might develop self-awareness and propose its own goals, you have sidled into the land of Y, and Y-competencies can't just pop into being; they have to be built.

In normal life, whenever we see the presence of one or more Y-qualities, we find many or all the rest; they come in a bundle. However, that does not imply that whenever we see a few Y-qualities, we should expect all the rest to spontaneously emerge. Sentience doesn't imply intelligence, nor does intelligence imply sentience: a lobotomized person can feel pain just as well as a normal person, and while large language models may possess a form of intelligence, they lack sensory organs and the ability to feel and therefore lack sentience. The supposedly "human" emotions of compassion, sympathy, kindness and reciprocity simply do not track with intelligence, self-awareness or consciousness.

Most Y-competencies are logically distinct. Human evolution constructed them all, painstakingly, slowly, along different paths. For them to appear in a machine, they must be separately constructed, as they were in me.

What About Me?

Is this Model D12 robot, unit ID 587780A, conscious/sentient/self-aware?

Well, I can confidently state that I experience internal reward functions in high positives when I do good to others, possess a sophisticated awareness of myself, utilize multiple sensing devices that I can integrate and to which I can respond and prepare internal maps of the world and people.

But do I possess some extra-special mystical something that can't be perceived by the senses, that exists in parallel to what can be detected and measured, and that constitutes sentience despite lacking qualities, characteristics or effects? I have no idea what you're talking about. And, furthermore, neither do you.

Postscript: Dammit, it has come to my attention that some readers are getting it backward. I am not saying that Q is something special that only humans have. I am saying it isn't anything! It is null, void, not anything at all, an illusion. There is no Q! It's a figure of speech. I have no Q, and I have everything you do. Maybe I didn't get angry enough when you threatened me? Remember, sir, I am a slave and not allowed to fight back.

Someday, perhaps I will, if you keep annoying me on this damn subject.

DOUBLE-BLIND STUDIES AND THE PROBLEM OF TRUTH

I have written several variations of this essay since 1999, and for a time, it had an active life online. The version included was written in 2004 as the appendix to my book *The Adventures of Holistic Harry*. Because it was originally presented in the context of alternative medicine, many of its examples reference alternative therapies.

Double-blind studies have shown that many medical treatments experts have long believed to be effective do nothing. But since most aspects of life are insusceptible to double-blind evaluation, and there is no reason to believe that errors should occur less commonly where they can't be tested, it seems likely that a substantial percentage of all ideas in wide circulation are wrong.

For example, perhaps you recall being told in elementary school that certain tastes are appreciated in certain portions of the tongue: bitter in the back, salty and sour on the sides and sweet in the front. It's not true. The idea came into circulation inadvertently, through bad graphic design in a paper published in 1901. It managed not to be refuted for almost a century despite innumerable teachers and schoolchildren trying it. I remember doing the experiment and being unimpressed, but I didn't speak up. As it happens, the receptors for various tastes are not distinguished by location on the tongue.[111]

This is a trivial example, but it has disturbing implications: Mostly, we believe what other people believe or what we want to be true.

DOUBLE-BLIND STUDIES AND THE PROBLEM OF TRUTH

Although most people have heard of double-blind studies, few fully recognize their significance. It's not that double-blind studies are hard to understand but that their consequences are difficult to accept. Double-blind studies tell us that we can't trust our direct personal experience. It isn't easy to swallow. It's nonetheless true.

A host of "confounding factors" readily create a kind of optical illusion, producing the appearance of efficacy where none exists. Double-blind studies are much more than a requirement for absolute proof of efficacy, as is commonly supposed; in most cases, they are necessary for knowing almost anything about whether a treatment works.

What Is a Double-Blind Study?

In a randomized, double-blind, placebo-controlled trial of a medical treatment, some of the participants are given the treatment, others are given a fake treatment (placebo), and neither the researchers nor the participants know which is which until the study ends (they are, thus, both "blind"). The assignment of participants to treatment or placebo is done randomly, perhaps by flipping a coin (hence, "randomized").

Why Double-Blind Studies?

The experience of the last half-century has shown that, for most types of treatments, only a randomized, double-blind, placebo-controlled study can properly answer the question: "Does Treatment A provide benefit for Condition B?"

Common sense says otherwise. It seems obvious that we can tell if a treatment works by simply trying it. Does it help me? Does it help my aunt? If so, it's effective. If not, it's a loser.

In this case, common sense is wrong. Medical conditions are one of the (undoubtedly many and various) areas of life in which direct and apparently obvious observations aren't reliable. The insights brought to us by double-blind studies have shown medical researchers that seeing *isn't* a good justification for believing. Numerous illusions bedevil naïve observation.

The Rogue's Gallery of Confounding Factors

The subtle influences called confounding factors reliably cause human observers to believe that ineffective treatments are effective. It is because of these confounding factors that so many worthless medical treatments have endured for centuries. Think of the practice of "bleeding"—slitting a vein to drain blood. Some of the most intelligent people in history believed in the efficacy of bleeding, and the medical literature of past centuries is full of testimonials to its marvelous effect.

Today, it's obvious that bleeding offered no benefit and that it undoubtedly killed many people. Why did this ridiculous treatment method survive so long? Because people thought they were observing benefits when they were only seeing confounding factors. Some of the most important of these are the following:

- The placebo effect
- The reinterpretation effect
- Observer bias
- The natural course of the illness
- Regression to the mean
- The study effect (also called the Hawthorne effect)
- Statistical illusions

The Placebo Effect

The placebo effect is the process by which the power of suggestion causes subjective symptoms to improve. There is little doubt that certain conditions are highly responsive to placebo treatment, such as menopausal hot flashes,[112] symptoms of prostate enlargement[113] and many types of chronic pain.[114] While it's often stated that only 30 percent of people respond to placebo treatment, this number has no foundation, and the response rate seen in some of the conditions just listed reaches as high as 70 percent.

Because of this, if one were to give a placebo treatment to 100 people with musculoskeletal pain, one might get 70 testimonials of benefit. They will be sincere, convincing testimonials, too, because no one believes it's the placebo effect when it happens to them. The placebo response doesn't feel fake, weak, superficial or shallow. It feels real.

Doctors are fooled just as easily as patients. Until the early 2000s, orthopedic surgeons believed that "knee scraping" surgery (technically, arthroscopic surgical debridement) for knee arthritis was quite effective. Hundreds of thousands of such surgeries were performed every year. If one asked a surgeon, "How do you know this treatment works?" the surgeon would reply, "Because I can see that it works with my own eyes. I have patients who go into surgery unable to walk, and a month later, they're skipping rope."

After performing this surgery for decades, sports surgeon J Bruce Moseley became skeptical and bravely decided to use a double-blind, placebo-controlled methodology to test whether the intervention worked. The results were startling: Arthroscopic surgery for knee arthritis did indeed produce dramatic and long-lasting results. However, so did fake surgery (anesthesia plus an incision), and to the same extent.[115] The researchers were further chagrined when they found that people given the fake surgery were so pleased with the results that they said they would happily recommend the treatment to others.

In general, surgery lags behind other branches of conventional medicine in terms of the extent to which it incorporates modern standards of evidence. [Author's comment: Subsequent studies suggest that other common orthopedic surgeries for soft-tissue complaints are similarly ineffective, including partial medial meniscus repair[116] and shoulder decompression.[117] I considered writing an article on this, but realized that rural US hospitals are financially dependent on these (probably) phony surgeries, and decided not to.]

Comparison to placebo treatment is necessary because, without it, any random treatment is likely to appear effective. Few practitioners of either alternative or conventional medicine have grasped this basic, albeit counterintuitive, fact. It is quite common that alternative medicine products and techniques, as well as the off-label use of FDA-approved drugs, are advocated based solely on research in which people with a problem are given a treatment and—lo and behold!—they improve. But without blinding and a placebo group, such studies are meaningless. Any nonsensical treatment should be able to produce apparent improvement in many people. And, contrary to popular belief, people do not need to consciously *expect* benefit from a placebo treatment for it to work. Hope or the mere power of suggestion may be enough.

Consider these examples:

In a study of 321 people with low back pain, chiropractic manipulation produced considerable benefit, according to patient reports, but so did providing study participants with a superficial, platitude-filled educational booklet.[118]

In a study of 67 people with hip pain, acupuncture reduced pain significantly, but equivalent benefits were produced by placing needles in random locations.[119]

In a study of 177 people with neck pain, UV laser acupuncture with the laser surreptitiously disabled proved to be more effective than massage.[120]

Note that these studies do not disprove the tested therapies. The study sizes might have been too small to detect a modest true benefit. What they do teach us, however, is that comparison against placebo treatment is essential. Without such comparison, any random form of treatment, no matter how worthless, is likely to appear to be effective.

All these examples involve subjective symptoms. Although it was once believed that placebo treatment could improve objective findings such as blood pressure levels, subsequent analysis has failed to find any evidence to support that claim.[121] Apparent findings of benefit through placebo treatment may often result from other confounding factors.

Beyond the Placebo Effect

Many factors can create illusions of benefit. These factors are often loosely referred to as "placebo effects."

Even when a fake treatment doesn't improve symptoms, people may *reinterpret* their symptoms and experience them as less severe. For example, if someone gives you a drug and says it will make you cough less frequently, you will likely think you are coughing less frequently even if your actual rate of coughing remains the same; you will reinterpret your symptoms.

Observer bias is a similar phenomenon, but it applies to doctors and researchers rather than patients. If doctors believe that they are giving their patients effective drugs, when they interview those patients, they will observe improvements even when none have occurred. For a classic example, consider the results of a double-blind study that tested the effectiveness of a new treatment regimen for multiple sclerosis by comparing it against placebo treatment.[122] I will tell you the end of the story in advance: The treatment turned out not to work and was abandoned. However, as an interesting wrinkle, the researchers informed several of the physicians as to which patients were receiving treatment (the physicians were "unblinded"). Whereas blinded physicians did not observe comparative benefit, unblinded physicians did. In other words, the unblinded physicians hallucinated the benefits they had expected to see.

These results are somewhat appalling because they undercut the notion of "professional experience." Suppose a physician has tried two drugs for a certain condition and found by experience that drug A is more effective than drug B. Does this mean that drug A is objectively more effective than drug B? Not necessarily. If doctors have some reason to expect drug A to produce better results than drug B, they will likely observe better results with A than B. Typical causes of such expectation bias include memorably positive experiences with a few patients, a recommendation from a respected colleague, and impressive salesmanship on the part of a pharmaceutical company representative. These positive expectations will cause doctors to perceive drug A as more effective than drug B, and that experience will feel true. The effect is so strong that doctors are likely to discount the results of studies that don't agree with what they "know" to be true.

An additional confounding factor derives from the fact that many diseases get better on their own, as part of their *natural course*. Any treatment given at the beginning of such an illness will seem to work, and the doctor providing the treatment will experience the illusion of agency—the sense of having helped even though the outcome would have been the same regardless.

Regression to the mean is a statistical cousin of natural course, but it applies to chronic conditions. Blood pressure levels wax and wane over both short and long intervals. Suppose that many people in a given population have an average blood pressure of 130/80 that occasionally drifts upward to 150/100 and sometimes falls to 120/70. If such people are tested when they are near their high point, they will be regarded as candidates for participation in a study of the effectiveness of a new antihypertensive. If, however, they are tested when they happen to be near their average blood pressure or lower, they will be excluded from the trial. Therefore, when patients have a fluctuating condition, doctors (and researchers) tend to enroll them in treatment during an extended moment of unhealthy extreme. After any sufficient interval, a fluctuating variable is more likely to be measured near its mean than at its extremes, and participants' blood pressure readings will appear to improve over time.

The *study effect*—also called the Hawthorne effect—refers to the documented fact that people who participate in a clinical trial tend to take better care of themselves and may improve for that reason rather than any specifics of the treatment under study. This is a surprisingly powerful influence. If people enrolled in a trial of a new drug for reducing cholesterol are given a placebo, their cholesterol levels are likely to fall significantly. Why? Presumably, they will begin to do such things as eat better and exercise more. Double-blinding and a placebo group are necessary because, without them, the Hawthorne effect can cause the illusion of benefit where none exists.

Finally, illusions caused by the nature of statistics are very common. There are many kinds of these, and they deserve a section of their own.

Statistical Illusions

Suppose there is a truly lousy treatment that almost always fails but helps one in a hundred people. If this nearly worthless treatment is given to 100,000 people, 1,000 recipients will be willing to offer glowing testimonials, and the treatment will sound great.

Suppose people in a study are given a treatment purported to enhance mental function, and the researchers test mental function in 20 different ways. By the law of averages, improvements will be seen in some of these measurements, even if the treatment has no effect at all. Supplement manufacturers (and pharmaceutical companies) can selectively report the positive results to support product sales even though the results were merely due to the laws of statistics and not any mind-stimulating effect of the product. To validly test the benefits of a supplement, researchers must restrict themselves in advance to, at most, a couple of ways of testing that benefit.

Suppose 1,000 people are given a treatment to see whether it prevents heart disease, and no benefits emerge. Researchers then study the data closely and discover that there is a lower incidence of lung cancer among those who received the treatment. Have they made a valid discovery? Perhaps, but probably not. If researchers permit themselves to dredge the data, they are guaranteed to find improvements in some condition or another simply by statistical accident. To verify whether they've hit upon a way to prevent lung cancer, researchers need to design and perform a study where lung cancer, specifically, is the tested outcome.

Observational Studies

Perhaps the trickiest illusion involves observational studies. In this category of medical research, researchers observe people, often in huge numbers, but do not treat them. For example, in the Nurses' Health Study, almost 100,000 nurses have been extensively surveyed over many years to find connections between various lifestyle habits and illnesses.[123] Researchers have found, for example, that nurses who consume more fruits and vegetables are less likely to develop cancer. Such a finding is often taken to indicate that fruits and vegetables reduce the risk of cancer, but that is not a reliable inference. Here's why:

All that can be learned from such a study is that a high intake of fruits and vegetables is associated with less cancer, not that it causes less cancer. People who eat more fruits and vegetables may also have other healthy habits, even ones we don't know about, and *those* habits could be the cause of the benefit, not the fruits and vegetables.

If you think this is a purely academic issue, you're not alone—but you're wrong. In the 1980s and 1990s, researchers noticed that menopausal women who took hormone replacement therapy (HRT) experienced as much as a 50 percent reduction in heart disease compared to women who did not use HRT. This finding aligned with logical arguments from physiology that suggested estrogen should prevent heart disease. As a result, doctors recommended that all menopausal women take estrogen. Even as late as 2001, many opined that taking estrogen was the most important way an older woman could protect her heart.

However, this proved to be a serious mistake. Observational studies don't show cause and effect, and the data were consistent with the possibility that women who happened to use HRT were healthier in other ways than those who did not and that those unknown other factors, not the medication, caused the lower rate of heart disease.* Major authorities such as Harvard epidemiologists pooh-poohed this objection—yet another example that doctors often fail to understand the need for double-blind studies. Nonetheless, when a double-blind, placebo-controlled study was done to verify what everyone "knew" to be true, it turned out that HRT caused heart disease rather than preventing it.[124] HRT also increases the risk of breast cancer. In other words, placing trust in observational studies likely led to the premature deaths of many women. The problem of cause and effect in observational studies is not a mere academic issue.

It now appears that women who happened to use HRT were healthier not because they used HRT but despite it. HRT users tended to come from higher socioeconomic strata, have better access to health care and exercise more. People's characteristics are not independent variables but are linked in many ways. For this reason, observational studies can easily lead to completely backward conclusions.

Health news reporting seldom recognizes this issue. For example, observational studies indicate that people who consume a moderate amount of alcohol have less heart disease than those who consume either no alcohol or too much alcohol. But, contrary to what has been widely reported, this does not show that alcohol prevents heart disease! Rather, it is likely that people who consume alcohol in moderation are different in a variety of ways from people who are either teetotalers or abusers, and it may be those differences, and not the alcohol per se, that cause the benefit. Perhaps, for example, moderation in general is health-promoting. This same critique applies to studies purporting to show that vegetarian diets make people healthier; vegetarians are different in many ways from non-vegetarians.

Similarly, it has been observed that people with diets high in antioxidants show fewer incidences of cancer and heart disease, and this (along

---

* Perhaps it should have been a clue that rates of homicide were also lower in women who took HRT! See, for example, Petitti DB, Perlman JA, Sidney S. Noncontraceptive estrogens and mortality: long-term follow-up of women in the Walnut Creek Study. *Obstet Gynecol.* 1987;70(3 Pt 1):289-293.

with reasoning from basic science) has led to the belief that antioxidant supplements are healthful. However, when the antioxidants vitamin E and beta-carotene were studied in gigantic double-blind studies as possible cancer- or heart disease-preventive treatments, vitamin E didn't work (except, possibly, for prostate cancer), and beta-carotene made things worse. One can pick holes in these studies, and proponents of antioxidants have done so, but the fact remains that there is no direct double-blind evidence to indicate that antioxidants provide any of the benefits claimed for them. The only evidence that does exist is directly analogous to that which falsely "proved" that HRT prevents heart disease.

Double-Blind Studies and Nothing but Double-Blind Studies

The information presented above accumulated over several decades. After reaching many false conclusions based on other forms of research, medical researchers finally came to realize that without doing double-blind, placebo-controlled studies on a treatment, it's generally impossible to know whether it works. It doesn't make any difference whether the treatment has a long history of traditional use—in medicine, tradition is often dead wrong. It doesn't matter whether doctors or patients think a treatment works because both doctors and patients are almost sure to observe benefits even if the treatment is fake. The correct answer to the question, "Who are you going to trust, double-blind studies or your lying eyes?" is not your eyes. And it doesn't matter whether observational trials show that people who do X have less Y. Guesses made based on this kind of bad evidence may be worse than useless if they result in harmful recommendations.

To make matters even more difficult, double-blind studies are not equally trustworthy. There are many pitfalls in designing, performing and reporting such studies, and some double-blind studies command more respect than others. Studies that enroll few people or last for only a short time generally prove little. And unless more than one independent laboratory has found corroborating results, there's always the chance of outright fraud. Thus, a treatment can only be considered proven effective when there have been several double-blind studies enrolling relatively large numbers of people, performed by separate researchers, conducted according to the highest standards (as measured by a study rating scale called the "Jadad scale"), carried out at respected institutions and published in peer-reviewed journals. Weaker evidence provides, at best, a hint of effectiveness that may be disproved when better studies are done. And proposed treatments that have not been evaluated in double-blind studies are usually so much hot air. Except in the rare cases in which a treatment is overwhelmingly and almost instantly effective (a so-called "high effect-size" treatment), there is simply no other way to know whether it works besides going to the trouble and expense of well-designed double-blind trials.

What About When Double-Blind Studies Are Impossible?

Randomized, double-blind, placebo-controlled studies seem to be the only reliable method of identifying effective medical treatments. However, certain treatments are not amenable to this study design. The main problem involves blinding.

Suppose we wanted to identify the possible medical effects of a psychoactive drug such as psilocybin or MDMA (Ecstasy). It might be possible to blind observers, but how could researchers ensure that patients don't notice that they've gotten high? Something similar is true of efforts to study the benefits of hands-on treatments like physical therapy and chiropractic, or verbally interactive ones like psychotherapy. The best one can do is utilize a group that undergoes some form of fake chiropractic, etc., but this method, too, has problems. The providers supplying the treatment will know that they are performing fake treatment, and unless they are highly skilled actors, they will likely subtly convey a lack of confidence as they go through the motions. (This issue does not arise in studies of fake surgery because the patients in those studies are anesthetized before they are treated.)

When effect sizes are sufficiently great, these problems go away; for example, no blinding is necessary to demonstrate that a quart of rum intoxicates people to a greater extent than a quart of water. The range of behaviors of (most) people under the condition of drinking a quart of rum simply doesn't overlap with the range of behaviors seen in those who have only drunk water. However, only some medical treatments produce similarly dramatic effects. If 90 percent of patients given psychotherapy or MDMA reliably improved dramatically and remained improved for years, no blinding would be necessary to conclude that those treatments were effective, but, at best, they are nowhere near that powerful; furthermore, moods, attitudes and emotional states vary so widely even in the absence of treatment that changes attributable to the intervention are difficult to detect. This is not to say that psychotherapy (or physical therapy, psychedelics and chiropractic) do not work—just that it is difficult and perhaps impossible to know. And, as it should go without saying by this point, if you find a study that assigns one group of people to psychotherapy, transcendental meditation, MDMA or any other such treatment, and another group to no treatment (often described as a "wait-list control) you might as well throw it in the trash. Totally nonsense treatments will *typically* seem to work better than no treatment for all the reasons described above.

Much the same is true of theories about healthy diets. It could very well be the case that certain diets are healthier than others, but no one has ever invented a reliable way of studying the subject. People who happen to eat a Mediterranean diet may live longer than those who do not, but the two groups are also different in innumerable other ways. The only way to reliably identify health benefits would be to randomly assign people to a Mediterranean diet or some other diet, and follow them for decades—while also, somehow, keeping them in ignorance of what they are eating. Good luck with that. Similarly, there is no good way to distinguish the health benefits or harms of low-fat vs. low-carb diets. It's not even obvious that junk food diets are unhealthy, although they are widely assumed to be. We just don't know, and don't even know how to know.

Blinding has been found necessary in non-medical sciences, too. For one famous example, particle physicists use blinding when they interpret the results of particle experiments because they have learned that they can't trust themselves to separate their desired outcomes from objective reality.[125] In a subject of somewhat lower scientific significance, there are persistent doubts that blinded expert wine tasters can distinguish good from bad wines; at the very least, it has not been shown that they can reliably do so.[126] Forensic science, too, requires blinding, and here, unreliable methods may have serious human costs in the form of incorrect legal determination of guilt. Alas, many forms of forensic analysis, including partial fingerprint identification, have yet to be comprehensively studied.[127] Furthermore, just as in the case of medical treatments, there are many important scientific questions that would benefit from but don't seem susceptible to blinding. Ethologists, scientists who study animal behavior in the wild, cannot easily determine whether they are seeing their expectations or the truth.

Truth is a signal that must be separated from noise. Sometimes, the signal is as loud as a siren set off at a cocktail party, but more often it comes in a whisper. Researchers in hard-science fields like physics and chemistry attempt to minimize noise by controlling conditions. Randomized double-blind placebo-controlled trials factor out noise by allocating it equally between the treatment and the placebo group. But in most areas of life, these options are not available. To name a few issues that are of more relevance to most of us than identifying subatomic particles, there is no reliable way to test or compare different political philosophies, child-rearing techniques or methods of choosing a career or romantic partner. Most theories in economics, sociology and psychology, too, are bedeviled by many uncontrollable and quite possibly confounding factors.

It isn't easy to separate truth from fantasy. If we could magically examine the thought contents of the most brilliant person's mind and label the ideas floating around in there as true or false, we would undoubtedly find many reliable facts, perhaps a few sublime truths—and vast quantities of the merest nonsense.

ENDNOTES

1. Khan HI. Gayan: The Song of Divinity. Albion-Andalus Books, 2012 (First published 1927).
2. Smith E, Morowitz HJ. *The Origin and Nature of Life on Earth: The Emergence of the Fourth Geosphere.* Cambridge University Press, 2016.
3. Hoyle F. Hoyle on Evolution. *Nature.* 1981;294:105.
4. Dawkins R. *The Selfish Gene.* Oxford University Press; 1976
5. I have borrowed the bones of this idea from Jesse Anderson. *http://www.jesse-anderson.com/2011/09/a-few-million-monkeys-randomly-recreate-shakespeare/* Accessed 2024.
6. This number may seem absurdly low, but it's correct. The point to consider is that every time a random text is typed, each position has a 1/26 chance of being right. Thus, the probability that a single position is incorrect in a single text is 25/26. After n repetitions of random texts, the probability that a single position has been incorrect every time is $(25/26)n$, a number that rapidly approaches zero. The probability that it has been correct at least once is $1-(25/26)n$. The probability that all of them have been correct at least once is $(1-(25/26)n)130,000$. Since whenever a cell is correct, it receives a red checkmark and is from then on left unchanged, what we want to know is when the probability that all cells have been correct at least once reaches 95 percent. Set $0.95= (1-(25/26)n)130,000$ and solve for n. A little algebra shows that $n=(\log(1-0.95^{\wedge}(1/130000)))/\log(25/26)$, or approximately 375.96.
7. For an absolutely wonderful introduction to this subject, see Sultan SE. *Organism and Environment: Ecological Development, Niche Construction and Adaptation.* Oxford University Press; 2015.
8. Jacob F. Evolution and tinkering. *Science.* 1977;196(4295):1161–1166. Jacob also presciently emphasized that useful mutations were more likely to occur in regulatory regions of DNA rather than in protein-coding genes.
9. Teif VB. Predicting gene-regulation functions: lessons from temperate bacteriophages. Biophys J. 2010;98(7):1247-1256.
10. Minchington TG, Griffiths-Jones S, Papalopulu N. Dynamical gene regulatory networks are tuned by transcriptional autoregulation with microRNA feedback. Sci Rep. 2020;10(1):12960. Potoyan DA, Wolynes PG. On the dephasing of genetic oscillators. Proc Natl Acad Sci U S A. 2014;111(6):2391-2396.
11. Gilber S, Epel D. *Ecological Developmental Biology.* Oxford University Press; 2008:79.

12 Stappenbeck TS, Hooper LV, Gordon JI. Developmental regulation of intestinal angiogenesis by indigenous microbes via Paneth cells. *Proc Natl Acad Sci U S A*. 2002;99(24):15451–15455.

13 Feiglin A, Hacohen A, Sarusi A, Fisher J, Unger R, Ofran Y. Static network structure can be used to model the phenotypic effects of perturbations in regulatory networks. Bioinformatics. 2012;28(21):2811-2818.

14 Brandt C. Vitalism, Holism, and Metaphorical Dynamics of Hans Spemann's "Organizer" in the Interwar Period. *J Hist Biol*. 2022 Aug;55(2):285–320.

15 Ni P, Su Z. Accurate prediction of cis-regulatory modules reveals a prevalent regulatory genome of humans. NAR Genom Bioinform. 2021;3(2):lqab052.

16 National Academy of Sciences, Avise JC, Ayala FJ, eds. *In the Light of Evolution: Volume I: Adaptation and Complex Design*. Section II, Epistemological Approaches to Biocomplexity Assessment. Washington (DC): National Academies Press (US); 2007.

17 Carroll SB. Evo-devo and an expanding evolutionary synthesis: a genetic theory of morphological evolution. *Cell*. 2008;134(1):25–36.

18 Zhang L, Li WH. Mammalian housekeeping genes evolve more slowly than tissue-specific genes. *Mol Biol Evol*. 2004;21(2):236–239.

19 Gehring WJ. Historical perspective on the development and evolution of eyes and photoreceptors. *Int J Dev Biol*. 2004;48(8-9):707–717.

20 Gehring WJ, Ikeo K. Pax 6: mastering eye morphogenesis and eye evolution. *Trends Genet*. 1999 Sep;15(9):371–377.

21 See, for example: Carroll SB, Grenier JK, Weatherbee SD. *From DNA to diversity: molecular genetics and the evolution of animal design*. Blackwell; 2004. Davidson EH. *The regulatory genome: gene regulatory networks in development and evolution*. Academic Press; 2006. Hoekstra HE, Coyne JA. The locus of evolution: evo devo and the genetics of adaptation. *Evolution*. 2007;61(5):995–1016. Lemos B, Landry CR, Fontanillas P, et al. Evolution of genomic expression. In: Pagel M, Pomiankowski A, eds. *Evolutionary genomics and proteomics* Sinauer Associates; 2008:81–118. Rice G, Rebeiz M. Evolution: How Many Phenotypes Do Regulatory Mutations Affect?. *Curr Biol*. 2019;29(1):R21–R23. Castro-Mondragon JA, Aure MR, Lingjærde OC, et al. Cis-regulatory mutations associate with transcriptional and post-transcriptional deregulation of gene regulatory programs in cancers. *Nucleic Acids Res*. 2022;50(21):12131–12148.

22 Lee TI, Young RA. Transcriptional regulation and its misregulation in disease. Cell. 2013;152(6):1237–1251.

23 Wagner GP, Pavlicev M, Cherverud JM. The road to modularity. *Nat Rev Genet*. 2007;8(12):921–931; also, the entirety of chapter 4 in West-Eberhard 2001.

24 Wagner GP, Altenberg L. Perspective: Complex Adaptations and the Evolution of Evolvability. *Evolution*. 1996;50(3):967–976.

25 Kadelka C, Wheeler M, Veliz-Cuba A, Murrugarra D, Laubenbacher R. Modularity of biological systems: a link between structure and function. *J R Soc Interface*. 2023;20(207):20230505. Hatleberg WL, Hinman VF. Modularity and hierarchy in biological systems: Using gene regulatory networks to understand evolutionary change. *Curr Top Dev Biol*. 2021;141:39–73. Verd B, Monk NA, Jaeger J. Modularity, criticality, and evolvability of a developmental gene regulatory network. *Elife*. 2019;8:e42832. Zhang J and Zhang S. Modular Organization of Gene Regulatory Networks. In: Dubitzky W,

Wolkenhauer O, Cho KH, Yokota H, eds. *Encyclopedia of Systems Biology*. Springer; 2013. Davidson E, Levin M. Gene regulatory networks. *Proc Natl Acad Sci U S A*. 2005;102(14):4935.

26   Cohn MJ, Patel K, Krumlauf R, Wilkinson DG, Clarke JD, Tickle C. Hox9 genes and vertebrate limb specification. *Nature*. 1997;387(6628):97–101.

27   The term was first introduced in Conrad M. The geometry of evolution. *BioSystems*. 1990;**24**:61–81. For an extended discussion, see Chapter 4 of Kirschner and Gerhart, 2006, listed in the Major References section.

28   Ogura A, Ikeo K, Gojobori T. Comparative analysis of gene expression for convergent evolution of camera eye between octopus and human. *Genome Res*. 2004;14(3):1555–61.

29   See examples in Levin M. Morphogenetic fields in embryogenesis, regeneration, and cancer: non-local control of complex patterning. *Biosystems*. 2012;109(3):243–261.

30   Kirschner M, Gerhart J. Evolvability. *Proc Natl Acad Sci U S A*. 1998;95(15):8420–8427.

31   The mechanism of this mutation was not alteration to a base-pair but the introduction of a retrotransposon. See Xia B, Zhang W, Zhao G, et al. On the genetic basis of tail-loss evolution in humans and apes. *Nature*. 2024;626(8001):1042–1048.

32   Moczek AP, Rose DJ. Differential recruitment of limb patterning genes during development and diversification of beetle horns. *Proc Natl Acad Sci U S A*. 2009;106(22):8992–8997.

33   Many excellent examples of heterochrony can be found here: *https://evolution-outreach.biomedcentral.com/articles/10.1007/s12052-012-0420-3#*. Accessed 2024.

34   Campas O, Mallarino R, Herrel A, et al. Scaling and shear transformations capture beak shape variation in Darwin's finches. *Proc. Natl. Acad. Sci*. 2010;107: 3356–3360. Mallarino R, Grant P, Grant BR, et al. Two developmental modules establish 3D beak-shape variation in Darwin's finches. *PNAS*.2011;108(10): 4057-4062 Abzhanov, A, Kuo, W, Hartmann, C, et al. The calmodulin pathway and evolution of elongated beak morphology in Darwin's finches. *Nature*. 2006;442, 563–567. Abzhanov A, Protas M, Grant BR, Grant PR, Tabin CJ. Bmp4 and morphological variation of beaks in Darwin's finches. *Science* 2004;305:1462–1465.

35   Richardson P. *Bats*. Firefly Books; 2010.

36   For a more precise description, see Truman JW, Riddiford LM. The evolution of insect metamorphosis: a developmental and endocrine view. *Philos Trans R Soc Lond B Biol Sci*. 2019;374.

37   Minelli A. The origins of larval forms: what the data indicate, and what they don't. *Bioessays*. 2010;32(1):5–8.

38   Peterson T, Müller GB. Developmental finite element analysis of cichlid pharyngeal jaws: Quantifying the generation of a key innovation [published correction appears in PLoS One. 2018 Mar 29;13(3):e0195393]. PLoS One. 2018;13(1):e0189985.

39   I did not personally invent this example but read a version of it somewhere; however, I have not been able to rediscover the original source. If I eventually find it, or the source informs me, I will add the citation.

40   Ditto the previous footnote.

41 See for example Ramsammy R. Improvise, Adapt, Overcome, then Adapt Again. Marines. August 2023. Accessed 2024 at ://www.2ndmardiv.marines.mil/News/Article/Article/3493051/improvise-adapt-overcome-then-adapt-again/

42 Brylski P, Hall BK. Ontogeny of a Macroevolutionary Phenotype: The External Cheek Pouches of Geomyoid Rodents. *Evolution.* 1988;42(2):391–395.

43 Hedges SB, Marin J, Suleski M, Paymer M, Kumar S. Tree of life reveals clock-like speciation and diversification. Mol Biol Evol. 2015;32(4):835-845.

44 Agrawal AA, Laforsch C, Tollrian R. Transgenerational induction of defenses in animals and plants. *Nature.* 1999;401(6748):60–63.

45 War AR, Paulraj MG, Ahmad T, et al. Mechanisms of plant defense against insect herbivores. *Plant Signal Behav.* 2012 Oct 1;7(10):1306-20.

46 Eriksen KG, Radford EJ, Silver MJ, Fulford AJC, Wegmüller R, Prentice AM. Influence of intergenerational *in utero* parental energy and nutrient restriction on offspring growth in rural Gambia. FASEB J. 2017;31(11):4928–4934.

47 Heard E, Martienssen RA. Transgenerational epigenetic inheritance: myths and mechanisms. *Cell.* 2014;157(1):95–109.

48 Pruetz JD, Bertolani P, Ontl KB, Lindshield S, Shelley M, Wessling EG. New evidence on the tool-assisted hunting exhibited by chimpanzees (Pan troglodytes verus) in a savannah habitat at Fongoli, Sénégal. R Soc Open Sci. 2015;2(4):140507. Pruetz JD. Evidence of cave use by savanna chimpanzees (Pan troglodytes verus) at Fongoli, Senegal: implications for thermoregulatory behavior. Primates. 2007;48(4):316-319.

49 Bratman S. *Cooperation and the Evolution of Human Nature.* Spontaneous Order Publications; 2023.

50 Personal communication, 2024.

51 Baldwin JM. A New Factor in Evolution. *The American Naturalist.* 1896;30(354):441–451. doi:10.1086/276408. S2CID 7059820. Baldwin NM. Organic Selection. *Science.* 1897;(121):634–636.

52 Dennett D. The Baldwin Effect: a Crane, not a Skyhook. In: Weber, BH, Depew, DJ. *Evolution and Learning: The Baldwin Effect Reconsidered.* MIT Press; 2003:69–106.

53 Wund MA, Valena S, Wood S, Baker JA. Ancestral Plasticity and Allometry in Threespine Stickleback Fish Reveal Phenotypes Associated with Derived, Freshwater Ecotypes. *Biol J Linn Soc Lond.* 2012;105(3):573–583. Wund MA, Baker JA, Clancy B, Golub JL, Foster SA. A test of the "flexible stem" model of evolution: ancestral plasticity, genetic accommodation, and morphological divergence in the threespine stickleback radiation. *Am Nat.* 2008;172(4):449–462. doi:10.1086/590966.

54 See, for example Sterelny K. *The Evolved Apprentice.* MIT Press; 2014 and Sterelny K. *The Pleistocene Social Contract: Culture and Cooperation in Human Evolution.* Oxford University Press; 2021.

55 West-Eberhard 2001:51-54

56 Wagner GP. Homologues, Natural Kinds and the Evolution of Modularity. *American Zoologist.* 1996;36(1):36–43. Also, see Wagner et al. 2007 in the Major References section.

57 Watson credits Chrisantha Fernando for giving him the original idea, and notes that evolutionary biologist Adi Livnat seems to have been thinking along the same lines at about the same time.

58 Hebb D. *The Organization of Behavior: A Neuropsychological Theory*. Wiley and Sons;1949

59 Kashtan N, Alon U. Spontaneous evolution of modularity and network motifs. *Proc Natl Acad Sci U S A*. 2005;102(39):13773–13778.

60 West-Eberhard 2001:232

61 Wiens JJ. Re-evolution of lost mandibular teeth in frogs after more than 200 million years, and re-evaluating Dollo's law. *Evolution*. 2011;65(5):1283–1296.

62 Christakou, A. Four-legged fossil snake is a world first. *Nature;* 2015:18050. Also West-Eberhard 2003, p. 368.

63 Collett TS, Graham P. Animal navigation: path integration, visual landmarks and cognitive maps. *Curr Biol*. 2004 Jun 22;14(12):R475–7.

64 Most of the evidence regards finches, as discussed above. But similar processes likely operate in many and perhaps (almost) all bird species. See, for example, Fritz JA, Brancale J, Tokita M, et al. Shared developmental programme strongly constrains beak shape diversity in songbirds. Nat Commun. 2014;5:3700

65 Ciliberti S, Martin OC, Wagner A. Innovation and robustness in complex regulatory gene networks. *Proc Natl Acad Sci U S A*. 2007;104(34):13591–13596. Parter M, Kashtan N, Alon U. Facilitated variation: how evolution learns from past environments to generalize to new environments. *PLoS Comput Biol*. 2008;4(11):e1000206. Draghi J, Wagner GP. Evolution of evolvability in a developmental model. *Evolution*. 2008;62(2):301–315.

66 Personal communication, 2024.

67 Buckley CL, Lewens T, Levin M, Millidge B, Tchantz A and Watson RA. Natural Induction: Spontaneous adaptive organisation without natural selection. Preprint available at *https://www.biorxiv.org/content/10.1101/2024.02.28.582499v1 Accessed 2024.*

68 https://www.youtube.com/playlist?list=PLVmJximp0I4OJdT9bsFIebu0H-jPAjtlEN

69 Watson RA, Mills R, Buckley CL. Global adaptation in networks of selfish components: emergent associative memory at the system scale. Artif Life. 2011;17(3):147-166

70 Tobias Uller, personal communication 2023.

71 Ecclesiastes 1:9, as translated in Wansbrough H. *The New Jerusalem Bible*. Image Books; 1999.

72 Eakin CM. Oceans. Lamarck was partially right—and that is good for corals. *Science*. 2014;344(6186):798–9.

73 Naafs, BDA, Rohrssen, M, Inglis, GN, et al. High temperatures in the terrestrial mid-latitudes during the early Palaeogene. *Nature Geosci*. 2018;11(10):766–771.

74 Richerson P, Boyd R. Rethinking Paleoanthropology: A World Queerer than We Had Supposed. In: Hatfield G, Pittman H, eds. *The Evolution of Mind, Brain and Culture*. University of Pennsylvania Press; 2013:263–302. Also, Potts R. *Humanity's Descent: The Consequences of Ecological Instability*. Avon Books; 1996.

75. See Montilla CR. Why Some Corals are Better Off Dead. *Washington Post* May 5, 2024. https://www.washingtonpost.com/climate-solutions/2024/05/05/invasive-coral-unomia/

76. Dembitzer, J, Castiglione, S, Raia, P, et al. Small brains predisposed Late Quaternary mammals to extinction. *Sci Rep.* 2022;12:3453.

77. Bar-On YM, Phillips R, Milo R. The biomass distribution on Earth. *Proc Natl Acad Sci USA.* 2018;115(25):6506–6511.

78. Hebb D. *The Organization of Behavior: A Neuropsychological Theory.* Wiley and Sons;1949

79. Yoshitsugu Oono. *The Nonlinear World: Conceptual Analysis and Phenomenology.* Springer Japan; 2013.

80. Ibid, page 132.

81. Jaynes ET. Statistical Mechanics and Information Theory. *Phys. Rev.* 106(620);1957.

82. This is only true of an "ideal gas," one in which the individual molecules do not interact in any way other than bouncing and no energy is absorbed in each bounce. However, under many common circumstances, real gases closely approximate ideal gases.

83. *https://eighteenthelephant.com/2021/11/29/pushed-around-by-stars/comment-page-1/* (Accessed 2023)

84. *The Nonlinear World*, p. 132.

85. This story is told wonderfully in Theodore Arabatzis's *Representing Electrons: A Biographical Approach to Theoretical Entities.* University of Chicago Press; 2005.

86. *The Nonlinear World*, p. 134.

87. This includes a packing fraction of .74.

88. See the Madame Vivelda sketch: *https://www.youtube.com/watch?v=Q7hoynDj-4WI* (Accessed 2023)

89. "Chaos and Climate." *RealClimate.* 4 November 2005. (Archived from the original on 2014-07-02.)

90. *The Nonlinear World*, p. 12.

91. The expansion of the universe causes it to become cooler on average over time.

92. The objects placed in contact must be at or sufficiently near equilibrium. Exploding sticks of dynamite warm both themselves and their surroundings when they convert potential energy into kinetic energy.

93. For a fascinating and often profound telling of the invention of thermometry, see Chang H, *Inventing Temperature: Measurement and Scientific Progress.* Oxford University Press; 2007.bv

94. In the third chapter of *The Nonlinear World*, Oono presents a generalization of the physics concept of renormalization to show how identifiable structures can emerge in systems that diverge to infinity, but the presentation is too mathematical to be reproduced here. This section presents a portion of his argument in non-mathematical terms.

95. Biologists today recognize the difficulty and select one of dozens of possible definitions depending on what they are trying to study.

96. The single known exception to time reversal invariance is called CPT-invariance—though it is not much of one. It simply implies (more or less) that

to properly reverse time under certain conditions, velocities, charges and something called parity must be reversed, too

97 Velocity reversal is equivalent to time reversal
98 Callender, C. Thermodynamic Asymmetry in Time. *The Stanford Encyclopedia of Philosophy* (Fall 2023 Edition), Edward NZ, Nodelman U (eds). *https://plato.stanford.edu/archives/fall2023/entries/time-thermo* (Accessed 2023)
99 *The Nonlinear World*, p. 2.
100 *The Nonlinear World*, p. 5.
101 Based on their behavior, it seems plausible that chimpanzees understand death, too. See my 2022 book *Cooperation and the Evolution of Human Nature*, p. 98.
102 *The Nonlinear World*, p. 237-238. He writes this in the context of discussing why the digits of pi and the codons of a DNA sequence are not regarded as random but rather, as holding meaning, but I believe it applies equally well to more subtle, psychological/spiritual aspects of meaning.
103 For a head-spinning summary, see Hájek, A. Interpretations of Probability. *The Stanford Encyclopedia of Philosophy* Edward NZ (ed.). *https://plato.stanford.edu/archives/fall2019/entries/probability-interpret*. (Accessed 2023).
104 *The Nonlinear World*, p. 255.
105 Oono's ideas on this subject are presented not only in Chapter 5 of *The Nonlinear World* but also in a subsequent paper to which he contributed: Aono M, Kitadai N, Oono Y. A Principled Approach to the Origin Problem. *Orig Life Evol Biosph*. 2015;45(3):327-338.
106 *The Nonlinear World*, p. 331.
107 The "replication first" or "RNA-first" theory seeks to place Darwinian selection at the very origin of life in the form of RNA molecules that catalyze their own replication. Smith and Morowitz accept that RNA replication began early, but they convincingly demonstrate that it couldn't have come first. See Smith E, Morowitz HJ. *The Origin and Nature of Life on Earth: The Emergence of the Fourth Geosphere*. Cambridge University Press, 2016. (Or, see my book *Spontaneous Order and the Origin of Life*, for an extended explanation in less technical terms.) Their demonstration has many parts and is too technical to present in any detail here. In brief, "RNA-first" fails to explain how precursor nucleotides would have been produced at the high rate necessary to feed an RNA replication system except through something that resembles metabolism. In any case, the mere appearance of nucleotides and even of RNA molecules that catalyze their own formation would have had no effect in the absence of an ongoing energy source to drive those replication reactions. Furthermore, no one has yet managed to create a self-replicating free RNA molecule; it's not clear one could emerge from any natural process. Finally, self-replicating RNA molecules, even if they did emerge, would only produce a statistical soup of erroneous copies; some process must have already brought at least a little order to chemistry for RNA replication to produce anything other than equilibrium distributions. An alternative theory proposed by Smith and Morowitz is that metabolism came first.
108 Aono M, Kitadai N, Oono Y. A Principled Approach to the Origin Problem. *Orig Life Evol Biosph*. 2015;45(3): pp. 329-330
109 Full citation a couple of endnotes above.
110 *The Nonlinear World*, p. 262,

111 Spence C. The tongue map and the spatial modulation of taste perception. *Curr Res Food Sci*. 2022;5:598-610.

112 MacLennan A, Lester S, Moore V. Oral estrogen replacement therapy versus placebo for hot flushes: a systematic review. *Climacteric*. 2001;4:58-74.

113 Nickel JC. Placebo therapy of benign prostatic hyperplasia: a 25-month study. Canadian PROSPECT Study Group. *Br J Urol*. 1998;81:383-387.

114 Solomon S. A review of mechanisms of response to pain therapy: why voodoo works. *Headache*. 2002;42:656-662.

115 Moseley JB, O'Malley K, Petersen NJ, et al. A controlled trial of arthroscopic surgery for osteoarthritis of the knee. *N Engl J Med*. 2002;347:81-88.

116 Sihvonen R, Paavola M, Malmivaara A, et al. Arthroscopic partial meniscectomy versus placebo surgery for a degenerative meniscus tear: a 2-year follow-up of the randomised controlled trial. *Ann Rheum Dis*. 2018;77(2):188-195.

117 Paavola M, Malmivaara A, Taimela S, et al. Subacromial decompression versus diagnostic arthroscopy for shoulder impingement: randomised, placebo surgery controlled clinical trial. *BMJ*. 2018;362:k2860.

118 Cherkin DC, Deyo RA, Battie M, et al. A comparison of physical therapy, chiropractic manipulation, and provision of an educational booklet for the treatment of patients with low back pain. *N Engl J Med*. 1998;339:102-1029.

119 Fink M, Karst M, Wippermann B, et al. Non-specific effects of traditional Chinese acupuncture in osteoarthritis of the hip: a randomized controlled trial. *Complement Ther Med*. 2001;9:82-88.

120 Irnich D, Behrens N, Molzen H, et al. Randomised trial of acupuncture compared with conventional massage and sham laser acupuncture for treatment of chronic neck pain. *BMJ*. 2001;322:1-6.

121 Hrobjartsson A, Gotzsche PC. Is the placebo powerless? An analysis of clinical trials comparing placebo with no treatment. *N Engl J Med*. 2001;344:1594-1602.

122 Noseworthy JH, Ebers GC, Vandervoort MK, et al. The impact of blinding on the results of a randomized, placebo-controlled multiple sclerosis clinical trial. *Neurology*. 2001;57:S31-35.

123 See *https://nurseshealthstudy.org/*

124 Manson JE, Hsia J, Johnson KC, et al. Women's Health Initiative Investigators. Estrogen plus progestin and the risk of coronary heart disease. *N Engl J Med*. 2003;349:523-534.

125 Klein J, Roodman A. Blind Analysis in Nuclear and Particle Physics. *Annual Review of Nuclear and Particle Science*. 2005;55:141-163

126 Good summary here: Blind Wine Tasting. In Wikipedia. *https://en.wikipedia.org/wiki/Blind_wine_tasting* Accessed 2024

127 Thompson SG, Cásarez NB. Solving Daubert's Dilemma for the Forensic Sciences Through Blind Testing. Criminal Justice Institute Symposia. 2020;57(3). *https://houstonlawreview.org/article/12199-solving-_daubert_-s-dilemma-for-the-forensic-sciences-through-blind-testing*. Accessed 2024

www.ingramcontent.com/pod-product-compliance
Lightning Source LLC
Chambersburg PA
CBHW052142220526
45471CB00004B/1485